Practical Android AI

A Hands-on Guide to Intelligent,Generative Apps at the Edge

by Zahidur Rahman Faisal

Practical Android AI

By Zahidur Rahman Faisal

Notice of Rights

Notice of Liability

Trademarks

ISBN: 978-1-950325-97-9

Table of Contents

Book License

Before You Begin

This section tells you a few things you need to know before you get started, such as what you'll need for hardware and software, where to find the project files for this book, and more.

What You Need

To follow the examples and exercises in this book, you will need a combination of software, hardware, and Google accounts. The items below cover the general prerequisites referenced throughout the chapters.

Development environment

- **Android Studio (Otter):** It's recommended to use the latest stable version of Android Studio and the SDK components (Android SDK Platform, emulator images). The minimum version of Android Studio supported by this book is **Otter**.
- **Android 6+:** The `minimumSdkVersion` for the projects is API 23, so you need a device or an emulator running API 23 or higher.
- **Kotlin 2.0 (or higher):** You need Kotlin **2.0 or higher** to run the sample projects. Use the Kotlin plugin bundled with the latest Android Studio releases, or the version indicated by the sample projects.

Hardware

Accounts & Cloud Services

- **Firebase Project with Billing Profile:** Basic generative AI features are free, but some advanced features like image generation may require setting up a billing profile.
- **Generative AI API keys:** For chapters that call generative models, you will need **Google/Firebase** API keys. Follow the instructions provided and set usage limits to control costs.
- **App Signing Key (Keystore):** For replicating deployment and delivery with AI bundles, you will need to create a keystore file (.jks or .keystore) and generate a private signing key using `keytool`.

Book Source Code & Forums

Where to Download the Materials for This Book

The materials for this book can be cloned or downloaded from the GitHub book materials repository:

- https://github.com/kodecocodes/aam-materials/tree/editions/1.0

Forums

We've also set up an official forum for the book at https://forums.kodeco.com/c/books/practical-android-ai. This is a great place to ask questions about the book or to submit any errors you may find.

To **Adam Farmer**, my mentor and colleague, who kindled my interest in Artificial Intelligence, and keeps fascinating me with his simple yet brilliant ideas!

To **Foyzul Karim**, my first tech-lead and AI enthusiast, who dared to hire a university student (me) and taught *"Hello World"* in Android – paving the path of my Mobile Engineering career.

And of course, to my family, friends, colleagues, and the team at Kodeco, who inspire me to keep learning and be better everyday.

— Zahidur Rahman Faisal

To my parents and family, for their unconditional love, constant encouragement, and for everything I am today — I am forever grateful.

To my love, **Niharika**, for your patience, belief in me, and unwavering support at every step of this journey.

— Prajwal S Belagavi

About the Authors

Zahidur Rahman Faisal is the author of this book — a digital nomad from Bangladesh with a passion for exploring new technologies and new places. He specializes in mobile innovation, with a particular focus on AI/ML, AR/VR and Wearables. Outside of programming, Faisal enjoys gaming, reading, hiking, or prototyping his latest creative ideas.

About the Editors

Prajwal S Belagavi is the final pass editor of this book. He resides in Bengaluru, India, and is an MBA graduate working at a leading credit card company. His interests span Android app development, Flutter, DeepTech, and Finance.He's a big time foodie, exploring breakfast places in Bengaluru. He loves trekking and flying his drone, and occasionally creates reels.

Introduction

Artificial Intelligence has transformed from a futuristic concept into an essential technology powering modern Android applications. Whether you're building productivity tools, creative apps, or intelligent assistants, AI and machine learning capabilities are now within reach for every Android developer.

This book guides you through the full landscape of AI and ML development on Android — from foundational concepts to practical implementation. You'll learn how to combine on-device intelligence for privacy and performance with cloud-based models for advanced capabilities. Through real-world examples and hands-on projects, you'll gain the skills needed to integrate machine intelligence into Android apps confidently and responsibly.

Whether you're exploring generative AI for the first time or expanding your expertise as an experienced mobile engineer, this book is designed to help you build intelligent, responsive, and ethical Android applications that users will love.

How to Read This Book

This book is structured so you can read it end-to-end for a cohesive learning path, but individual sections are also designed to stand alone so you can jump to areas that match your needs.

- **If you are new to AI on Android:** Read the chapters in order to build a strong conceptual foundation before diving into implementation.
- **If you are experienced and want targeted guidance:** Jump directly to the frameworks or techniques relevant to your project — on-device ML, generative AI, cloud intelligence, or interactive assistants.
- **Hands-on readers:** Follow the examples in each chapter and reproduce the small projects. Each section includes recommended sample apps and recipes that make it easier to try ideas quickly.
- **Team Leads and Architects:** Use the architectural, privacy, and productionization discussions throughout the book to guide real-world engineering decisions.

This book is split into three main sections. Each section consists of multiple chapters. Each chapter provides a concise overview on a topic and points to key examples and exercises.

Section I: Foundations of AI on Android

Artificial Intelligence is reshaping the Android ecosystem faster than any platform shift before it. Just a few years ago, integrating AI into a mobile app required deep ML expertise, heavy infrastructure, and complex custom models. Today, however, Google's AI stack — from Gemini to on-device engines like AICore and ML Kit — has made intelligent features accessible to every Android developer.

This first section gives you the foundational understanding you need before building AI-powered apps. You'll explore how AI is transforming Android, how to use AI tools to accelerate development, and how to get started with generative AI in your applications.

In this section, you'll learn:

- The evolving landscape of Android AI and the forces driving this shift.
- How **on-device** and **cloud-based** AI differ — and when to use each.

- How to use AI-assisted developer workflows, from smart code completion to **Gemini in Android Studio**, Gemini Agent Mode, and AI-driven debugging.
- Essential generative AI concepts: prompts, context, tokens, and model behavior.

Through these three chapters, you'll build a strong conceptual and practical foundation — preparing you for the deeper, more advanced AI features explored later in the book.

Section II: Building Core Intelligence

By now, you've explored the foundations of AI on Android and learned how today's ecosystem makes it possible to build smarter, more adaptive apps.

This section shifts the focus from concepts to practical, hands-on implementation. Here, you'll work directly with the core Android AI toolset — the frameworks and runtimes that power both on-device and cloud-based intelligence. You'll learn how to choose the right approach for your use case, integrate AI smoothly into your app's architecture, and deliver real machine intelligence that feels fast, reliable, and user-friendly.

Across these three chapters, you'll explore:

- **ML Kit for On-Device Intelligence**: Build document scanners, text extractors, and vision-powered features that run privately and instantly on the user's device.
- **MediaPipe for Custom ML**: Create your own ML pipelines and even run lightweight LLMs on-device, unlocking flexible, real-time AI experiences tailored to your app.
- **Firebase AI Logic for Cloud Power**: Offload complex or high-quality generative tasks to Gemini in the cloud, blending device and server intelligence into a hybrid architecture.

In this process, you'll have a solid command of the tools needed to build production-quality AI features — from vision to text to generative models.

Section III: Advanced Integration, Distribution, and Responsible AI

By this point in your journey, you've explored both the fundamentals of Android AI and the core tools that power intelligent features. Now it's time to move beyond implementation and into the realities of shipping, scaling, and sustaining AI features in production.

In this section, you'll learn:

- How to **package and deliver** on-device ML and GenAI models through the Play ecosystem, enabling dynamic model updates, optimized distribution, and reduced app sizes.
- How to build real-time, multimodal, assistant-like experiences with **Gemini Live**, including streaming audio, session management, and function calling for interactive agents.
- How to **design AI responsibly**, incorporating fairness, transparency, safety, and user control into every part of your app — from data flow to UI.
- How to prepare your AI features for production, covering monitoring, model rollback, budgeting, privacy constraints, and long-term sustainability.
- What the future of Android AI looks like, and how developers can adapt to the rapidly evolving ecosystem.

Across these final chapters, you will not only deepen your technical expertise but also gain the strategic perspective needed to build AI-powered Android apps that scale — ethically, safely, and confidently.

We hope you're ready to jump in and enjoy getting to know the power of AI in Android!

Section I: Foundations of AI on Android

Artificial Intelligence is reshaping the Android ecosystem faster than any platform shift before it. Just a few years ago, integrating AI into a mobile app required deep ML expertise, heavy infrastructure, and complex custom models. Today, however, Google's AI stack — from Gemini to on-device engines like AICore and ML Kit — has made intelligent features accessible to every Android developer.

This first section gives you the foundational understanding you need before building AI-powered apps. You'll explore how AI is transforming Android, how to use AI tools to accelerate development, and how to get started with generative AI in your applications.

In this section, you'll learn:

- The evolving landscape of Android AI and the forces driving this shift.
- How on-device and cloud-based AI differ — and when to use each.
- How to use AI-assisted developer workflows, from smart code completion to Gemini in Android Studio, Gemini Agent Mode, and AI-driven debugging.
- Essential generative AI concepts: prompts, context, tokens, and model behavior.

Through these three chapters, you'll build a strong conceptual and practical foundation — preparing you for the deeper, more advanced AI features explored later in the book.

Chapter 1: The AI-Powered Android Landscape

Zahidur Rahman Faisal

Artificial Intelligence (AI) is a field of computer science focused on creating machines that can perform tasks traditionally requiring human intelligence. These tasks include learning, reasoning, problem-solving, perception, understanding language, and making decisions. AI systems are designed to analyze data, identify patterns, and adapt their behavior to achieve specific goals, often mimicking cognitive functions found in humans. It's a broad field encompassing various sub-disciplines like machine learning (ML), natural language processing (NLP), and computer vision.

AI is fundamentally reshaping the Android ecosystem, moving beyond simple automation to enable highly intelligent, personalized, and proactive user experiences. This transformation is driven by rapid advancements in generative AI and the deep integration of AI capabilities into both the Android operating system and the developer toolchain.

At Google I/O 2025, Google emphasized the widespread integration of Gemini models, the emergence of agentic applications, and a robust suite of developer tools aimed at democratizing AI development for Android. These innovations lower the barrier for developers to incorporate AI seamlessly into mobile applications.

Agentic AI: Beyond Reactive Systems

Agentic AI refers to an AI system capable of making autonomous decisions based on its current assessment and past performance to accomplish a task, operating with limited or no human oversight. It combines multiple AI models in an orchestrated, integrated way, allowing a program to act autonomously within a broader environment. While generative AI enables content creation and decision-making, agentic AI builds upon it to enable autonomous action.

Agentic AI is considered the "third wave" of AI development, moving beyond the initial burst of technologies like recommendation engines (first wave) and generative AI (second wave). Unlike traditional AI, agentic AI is not confined to a simple input/output model. It has the independence to seek information, connect with other tools, and perform complex, multi-step actions towards a goal.

A helpful analogy illustrates this difference: if generative AI is a *Toolkit to DIY fix a leaky sink*, then an agentic AI system is like a *General Contractor* who can direct the plumber, coordinate with electricians, and investigate related damage. This highlights its orchestration and proactive problem-solving capabilities.

Key Characteristics

The distinction between traditional AI (input/output) and agentic AI (autonomous, goal-oriented, multi-step planning) signifies a fundamental shift in how users and developers will interact with AI. It moves from "asking a question and getting an answer" to "delegating a goal and letting the AI figure out the steps".

This transition implies a higher level of trust and sophistication required from AI systems, and a change in user expectations. For developers, it means designing applications that expose APIs and tools for agents to leverage, and building interfaces for monitoring and guiding autonomous AI workflows rather than just initiating single tasks. The "General Contractor" analogy emphasizes that agentic AI is not just about generating content but about orchestrating multiple AI models, tools, and even human interactions to achieve a complex objective. This represents a higher-order capability, driven by the increasing complexity of user problems and the proliferation of specialized AI models, which necessitate an AI that can intelligently combine and manage these diverse resources.

Agentic AI systems possess several defining characteristics:

- **Autonomy:** They can operate independently without continuous human intervention.
- **Goal-Oriented Decision Making:** Unlike traditional AI that primarily responds to inputs, agentic AI sets goals and actively pursues them.
- **Multi-Step Planning and Execution:** They can break down large, complex tasks into smaller, manageable steps and execute them sequentially. An agentic system can assess its progress, add new steps, or seek help from humans or other AI systems when needed.
- **Adaptability and Learning:** Agentic AI can improve over time through techniques like reinforcement learning and self-improvement.
- **Collaboration:** These systems can work effectively with other AI agents or humans to achieve goals.

The Impact of Agentic AI on Mobile Development Workflows

The concept of agentic applications represents a significant evolution in AI, moving beyond reactive, input-output systems to proactive, goal-oriented entities.

Agentic AI is poised to revolutionize mobile app development by addressing unique challenges such as device fragmentation and performance optimization, moving beyond traditional collaboration tools. The way agentic AI actively understands and responds to these challenges implies a shift—from developers merely using tools to delegating tasks to intelligent agents that grasp broader context and objectives.

This could fundamentally redefine roles within a mobile development team. For example, a product manager might interact directly with an agentic AI to refine feature priorities based on real-time user data, rather than relying solely on manual analysis and testing. The developer's role evolves into one that is more supervisory and creatively strategic, focusing on higher-order problem-solving.

Here's how agentic AI changing the scene for mobile development:

Coding Assistance

AI agents specialize in generating code snippets from natural language prompts, reducing manual effort. They can also find and fix "code smells", optimize code for performance and size, and suggest enhancements. Tools like GitHub Copilot, powered by OpenAI's Codex, are examples of AI-powered pair programmers.

Designing Apps

Context-aware AI agents analyze user interactions, environmental factors, and device capabilities to create intuitive experiences. They assist with layout creation, color selection, typography, and user flow optimization, while automating tasks like A/B testing. Platforms like Uizard convert sketches into functional UI components, accelerating design cycles.

Testing

With mobile applications growing in complexity, agentic AI introduces smart testing capabilities, including:

- Identify potential failure points before they occur.
- Intelligent test case prioritization for high-risk areas.
- Automated root cause analysis to pinpoint defects.

This represents a shift from reactive bug fixing to preventative quality assurance, resulting in higher-quality apps, fewer critical bugs, and improved user satisfaction which directly influences app store ratings and user retention.

Quality Assurance

Agentic AI enhances deployment workflows through:

- Code quality and security vulnerability detection before app store submission.
- Comprehensive test coverage generation.
- Cross-platform compatibility verification.

This predictive approach reduces costly emergency patches and ensures stable releases, maintaining user trust and developer credibility.

Streamlining Development Lifecycle

Agentic AI integrates tools across the development pipeline, eliminating handoff frictions between teams. It pulls together disparate data sources — design mockups, code repositories, feedback, and performance analytics — into a single ecosystem. With this, product managers can directly view how design decisions affect app performance metrics, helping dismantle traditional silos.

Agentic AI transforms product planning from opinion-based guesswork into a data-driven strategy. It can analyze massive data volumes, predict potential outcomes, and identify actual feature priorities based on usage patterns and revenue impact. This helps mobile teams align their product strategy with both user needs and business objectives. This creates a more integrated, agile and efficient development pipeline.

AI agents can automate task assignments, detect bottlenecks, and streamline processes, ensuring teams work more efficiently by prioritizing tasks and improving collaboration. "AI-Driven Mobile App Creation" platforms like replit (https://replit.com) suggests that agentic AI will lower the barrier to entry for app development. This implies that agentic AI will not just make professional developers more efficient, but it will also empower non-technical users or small businesses to create functional Android applications with minimal or no coding, potentially expanding the overall Android app ecosystem significantly.

A New Era for Android AI

Google underscored AI as the central pillar of it's strategy, with a particular focus on its integration across the Android platform, during Google I/O 2025. The sheer volume and depth of AI-related announcements highlight a strategic shift towards making AI ubiquitous and deeply functional for both users and developers.

The conference demonstrated a strategic pivot, re-architecting Google's product ecosystem around a unified AI foundation. This approach suggests that future Android development will inherently involve interacting with Gemini's capabilities, whether explicitly through APIs or implicitly through system-level enhancements. This pervasive integration means AI will become a fundamental expectation for Android users, pushing developers to adopt AI-first design principles. It also implies that Google aims to control the core AI layer, potentially making it harder for other AI models to gain significant traction within the Android ecosystem at a system level.

Gemini Across Google and Android

Gemini was the undeniable highlight of I/O 2025, being mentioned 147 times in the keynote and integrated across the entire Google product ecosystem! This signifies Google's strategy to position Gemini as the foundational AI layer across all offerings, including Android.

Gemini Integrated Google Search

Google Search is evolving with widespread AI Overviews, which are Gemini-powered summaries appearing at the top of search results, now reaching over 1.5 billion users monthly across 200 countries. The introduction of **AI Mode**, powered by Gemini 2.5, reimagines the search experience by handling complex queries, generating custom charts and graphics, and supporting follow-up interactions. This shift directly impacts how Android users consume and interact with information, making the experience more conversational and intelligent.

Gemini Powered OS

The statement that "it's no longer all about the OS" reflects a paradigm shift. While Android 16's **Material 3 Expressive** provides the aesthetic, but AI serves as the underlying intelligence. This evolution is driven by Google's investment in technologies like Gemini 2.5, Deep Think, and Veo 3. The effect is that AI is becoming the primary driver of innovation and user value, with the operating system acting as the conduit for these AI-powered experiences. This will likely lead to a re-evaluation of what constitutes a "core" Android feature versus an "AI-enhanced" feature.

Gemini Live

Gemini Live now incorporates **Project Astra**'s camera and screen-sharing capabilities into the Gemini app, rolling out on both Android and iOS. This transforms Gemini into an interactive, multimodal assistant that understands visual and contextual input that offers real-time help from a phone's camera feed.

Updates to Gemini 2.5 models were also announced, including an enhanced reasoning mode called **Deep Think** for Gemini 2.5 Pro. This provides more sophisticated capabilities for complex tasks, such as turning photo grids into 3D spheres with narration or performing on-the-fly language changes with text-to-speech features.

Gemini in Android Wear and Android Auto

Gemini's presence is expanding beyond smartphones to TVs, watches, and cars, allowing users to get things done on the go across their Android device ecosystem. This indicates a holistic AI strategy where the user's AI assistant is seamlessly available across all their connected Android-powered devices, offering consistent experiences.

The expansion of Gemini across various devices points to AI as the unifying element for a truly interconnected, multi-device Android experience. AI is positioned not just for novel features but as a critical utility for seamless interaction across diverse form factors and for bridging real-world communication gaps. The focus on hardwares and smart devices highlights AI's role in breaking down communication barriers and enhancing accessibility. This suggests that Android developers should consider designing AI-powered features that extend beyond the traditional smartphone screen.

Generative AI for Creative Applications

Among the standout announcements was **Veo 3**, a powerful video generation model that can create cinematic videos with native audio, including music, singing, and lip-syncing! This opens up immense possibilities for Android applications in content creation, video editing, and entertainment, allowing developers to integrate advanced media generation capabilities.

Google also introduced **Flow**, an AI filmmaking tool building on VideoFX, with features like camera motion, perspective control, and the ability to edit or extend generated video shots. Alongside this, **Imagen 4** (for image generation) and **Lyria 2** (for music) offer Android developers state-of-the-art media generation tools, enabling the creation of new app categories in creative and visual experiences.

The announcement of tools like **Veo 3**, **Flow**, and **Imagen 4** for developers, artists, designers, film makers and content creators signifies a push to make cutting-edge generative AI accessible. This is not just about Google's own applications; it is about empowering the broader developer community to build with these powerful tools. This will likely lead to an explosion of creative and multimedia-focused Android applications leveraging these generative capabilities.

It also places a greater responsibility on developers to consider the ethical implications, such as the potential for deepfakes or misinformation. To address concerns around AI-generated content, Google introduced **SynthID** detector, which can scan images, audio, video, or text for SynthID watermarks, indicating AI origin. This is crucial for maintaining trust and transparency in an AI-driven content landscape, especially for applications dealing with media creation and consumption on Android.

Enhancing User Experiences with AI

Beyond generative media, AI is being woven into everyday user interactions. Features like **AI Overviews** and **AI Mode** in Search aim to provide more intuitive, conversational, and context-rich access to information. For example, Google Meet now offers real-time translation that can match a speaker's tone and cadence, as demonstrated with Spanish-to-English conversion. This has significant implications for communication apps on Android, enabling seamless multilingual interactions.

With user permission, Gemini models can leverage personal context from Google apps like Gmail and Drive to suggest personalized responses that match the user's tone and style. This elevates AI from providing generic suggestions to offering context-aware, intelligent productivity support on Android devices.

AI Tools for Developers

Google I/O 2025 also provided Android developers with a concrete toolkit to integrate these new AI capabilities into their applications. This goes far beyond code generation - it enables:

- Automation of repetitive tasks.
- Intelligent assistance with complex development challenges.
- Streamlining of the entire development lifecycle, from design to deployment.

The AI powered workflow aims to significantly boost developer productivity and reduce time-to-market for Android applications. It also suggests a future where developers will increasingly rely on AI companions, shifting their role from purely coding to more high-level problem-solving, architectural design, and AI orchestration.

Significant Enhancements to Gemini in Android Studio (Including Agent Mode)

Gemini in Android Studio is positioned as an AI-powered coding companion that understands natural language and designed to boost developer productivity. It assists with a wide range of tasks, such as:

- Code completion and transformations.
- Maintaining naming conventions and style guides.
- Writing code documentation and commit messages.
- Creating Jetpack Compose previews.
- Building UI from design images.
- Analyzing crash reports and logs.
- Writing unit and integration tests.

This integration aims to accelerate development cycles and improve app quality.

Agent Mode

Agent Mode is a significant experimental capability feature that brings agentic AI capabilities into the IDE, allowing developers to describe complex developmental goals in natural language, such as generating unit tests or performing intricate code refactoring. The agent then formulates and executes a multi-step plan that can span multiple project files, utilizing various IDE tools.

Developers maintain full control, reviewing and approving changes at each step, with an optional “auto-approve” feature. This represents a shift towards more autonomous, goal-oriented assistance within the IDE.

Image to Code

The **Image to Code** feature aims to convert design mockups into functional Compose UI code, bridging the gap between UX design and development. This directly streamlines the UI creation process. Additional tools like **Journeys** and the **Version Upgrade Agent** introduce new agentic AI workflows for guided app building, testing, and codebase upgrades — further enhancing development speed and accuracy.

Model Context Protocol (MCP)

Gemini in Android Studio can also interact with external tools via the **Model Context Protocol (MCP)**, an open standard pioneered by Anthropic PBC. This provides a standardized way for AI models to use tools and communicate with other AI agents, extending its capabilities beyond the immediate IDE environment. The Model Context Protocol further hints at a future where these agents can interact with other tools and services, creating highly automated and interconnected development pipelines.

With a Gemini API key, Agent Mode's context window can be extended to 1 million tokens using Gemini 2.5 Pro, enabling the AI to reason across larger, more complex codebases and deliver more accurate and useful responses.

The ability of Agent Mode to "formulate an execution plan that can encompass multiple project files" and "execute under your direction" signifies a qualitative leap in AI assistance. This heralds a future where development environments become more intelligent, capable of understanding high-level goals and autonomously performing complex tasks, with developers acting more as supervisors and refiners.

Firebase AI Logic for Harnessing Powerful Models

Firebase AI Logic was introduced to help developers bring the power of generative AI to their mobile and web apps. It provides a streamlined path for client-side apps to interact with robust cloud AI services. This allows developers to harness more powerful cloud-based models like **Gemini Pro**, **Gemini Flash**, and **Imagen** for complex use cases, such as image generation or processing extensive multimodal data.

Firebase AI Logic enables applications to understand and process various modalities — text, images, audio, video, and documents — and then generate text, code, or media.

Introduction of New ML Kit GenAI APIs

ML Kit, Google's mobile SDK for on-device machine learning, announced new GenAI APIs. These APIs leverage **Gemini Nano** for common on-device tasks, providing out-of-the-box quality for features such as summarization, proofreading, rewriting, and image description. This is significant because it simplifies the integration of generative AI directly onto the device, reducing the complexity for developers.

The availability of ML Kit GenAI APIs for on-device Gemini Nano alongside Firebase AI Logic for cloud-based Gemini Pro / Flash / Imagen demonstrates a tiered approach. Developers can choose the right AI model and deployment strategy based on their app's specific needs for latency, privacy, and computational power.

The diverse requirements of mobile applications, such as real-time on-device processing versus complex cloud inference, necessitate a flexible ecosystem. This ensures broader applicability of AI across different app types and hardware capabilities.

AI's Expansion into Android XR and Wear OS 6

Google put another spotlight on **Android XR** - its platform for augmented, mixed, and virtual reality - aiming to replicate Android's success in smartphones within the XR space. A second Android XR device was announced, and a live translation feature for Android XR was demonstrated with a smart glasses prototype. This signifies a future where Android AI extends to immersive computing, enabling new forms of interaction and utility in AR/VR headsets and smart glasses.

With the launch of **Wear OS 6**, featuring **Material 3 Expressive**, wearables now offer personalized, motion-rich UI experiences. While not explicitly AI-driven in the provided information, the "expressive design" aims for more engaging and intuitive experiences, which can be further enhanced by underlying AI for personalization and adaptive interfaces.

The Risks and Challenges in Android AI

The transition from passive chatbots to agentic AI—models capable of taking independent action within the Android ecosystem—represents a paradigm shift. However, granting an AI the autonomy to navigate apps, manage credentials, and execute intents introduces a complex matrix of risks.

Below is an excerpt regarding the critical challenges you, as a developer may face when integrating agentic workflows on mobile devices.

Contextual Ambiguity

Android is a noisy environment. Notifications pop up, layouts change based on screen size, and UI elements often lack semantic labels. An agent relying on computer vision or view-hierarchy analysis can easily lose context.

For example, asking an agent to "Reply to the last message" is ambiguous if three different messaging apps have active notifications. Without rigid context-awareness, the agent may execute the command in the wrong domain.

The On-Device Resource Tax

Running capable agentic models locally (on-device) to preserve privacy imposes severe physical constraints on Android hardware.

- **Battery Life:** Continuous inference drains power significantly faster than standard app usage.
- **RAM:** Large context windows (needed to remember app states) compete with the OS for limited memory, risking background app kills.
- **Thermals:** Sustained NPU/GPU usage leads to thermal throttling, causing the UI to stutter or the device to dim.

When Hallucinations Become Actions

In a standard LLM chat, a "hallucination" (factual error) is misleading. In an agentic context, a hallucination can be destructive. The cost of an error scales with the autonomy of the agent.

For example, an agent misinterpreting a command:

- **Chatbot:** "I'm sorry, I made up a fact about history."
- **Agent:** "I have deleted the wrong photo album." or "I sent the email to the wrong recipient."

The stochastic nature of generative AI clashes with the deterministic requirement of mobile commands (Intents). Ensuring an agent calls the expected function with exact arguments still is a massive engineering hurdle!

The Breakdown of the Sandbox

For over a decade, Android security has relied on 'sandboxing' — isolating apps so one cannot see the data of another without explicit permission. Agentic AI, by definition, requires breaking these walls. To be useful, an agent must "see" the screen, interact with third-party apps, and pass data between them. That induce two major risks:

1. **Data Leakage:** An agent acting as a "middleman" between a banking app and a budgeting tool creates a new vector for data interception.
2. **Permission Fatigue:** To function, agents often require Accessibility Services or broad-scope permissions. This desensitizes users to granting high-level access, potentially opening doors for malware masquerading as "AI helpers".

Future Outlook

Latest developments in AI within the Android ecosystem signal a future where AI is not merely a feature but an intrinsic layer of the Android experience, from system-level intelligence to sophisticated developer workflows. This strategic shift positions AI as a fundamental expectation for Android users, influencing how applications are designed and how users interact with their devices.

Google I/O 2025 unequivocally demonstrated that AI is no longer an optional add-on but the core intelligence driving the future of Android. From pervasive Gemini integration across user experiences to sophisticated agentic tools for developers, AI is reshaping how applications are built and how users interact with their devices. The strategic balance between on-device and cloud AI, often favoring a hybrid approach, underscores a commitment to performance, privacy, and scalability.

For Android developers, this AI-powered landscape presents immense opportunities! The availability of tiered Gemini models, simplified ML Kit APIs, seamless Firebase AI Logic, and intelligent Android Studio capabilities empowers them to create more intuitive, personalized, and powerful applications. Developers must adapt by embracing AI-first design principles, understanding the nuances of on-device versus cloud deployment, and leveraging agentic tools to accelerate their workflows. The shift means focusing more on defining complex problems and orchestrating AI solutions, rather than primarily writing lines of code.

Anticipated future trends in the AI-powered Android landscape include continued advancements in multimodal AI, deeper integration of agentic capabilities across the operating system and developer tools, and a relentless focus on optimizing AI for edge devices.

Ethical considerations - including bias mitigation and data privacy - will remain paramount. As AI becomes more autonomous, the role of human oversight and the development of robust, transparent AI systems will be critical. The competition between Google and other technology giants, such as Apple, in delivering best-in-class mobile AI experiences will continue to drive rapid innovation, ultimately benefiting Android users with increasingly intelligent and helpful applications.

Conclusion

This book serves as your comprehensive guide to integrating the next wave of intelligence into your Android apps. We begin by exploring Agentic AI, defining what it means for your apps to transition from reactive tools to autonomous digital assistants capable of executing complex, cross-app workflows. You'll master the developer side of this shift—delving into the powerful AI-powered productivity features in Android Studio and the various AI agents in Android that you can leverage today.

The later chapters provide a deep dive into the technical stacks for custom intelligence. You'll compare and contrast the different approaches for deploying models, starting with accessible on-device ML using ML Kit for common tasks and then moving to custom ML with MediaPipe for building complex, real-time media processing pipelines. To conquer tasks too complex or costly for the device, the book will cover harnessing cloud power with Firebase AI Logic.

You'll also learn to bridge the gap between AI and UX by learning to build highly interactive apps with Gemini Live, along with the strategies to deploy apps with larger AI models using the **Play for on-device AI** APIs.

The book concludes with critical insights into best practices, ethics, and the future of Agentic AI in Android, ensuring you build powerful, responsible, and user-centric applications ready for the intelligent future.

So buckle up your seatbelt and get ready to dive in!

Chapter 2: AI-Powered Developer Productivity with Android Studio & Gemini

Zahidur Rahman Faisal

Android Studio is being transformed into an AI-powered development environment, with Gemini deeply integrated to enhance developer productivity throughout the entire software development lifecycle. The sheer breadth of Gemini's integration into Android Studio – from basic code completion to advanced crash report analysis and multi-file refactoring via Agent Mode – indicates Google's ambition to make Android Studio the definitive AI-native development environment. It is not just adding AI features; it is embedding AI into the core fabric of the IDE. This deep integration will likely set a new standard for developer tools, pushing other IDEs and platforms to follow suit. It aims to create an unparalleled developer experience for Android, potentially attracting more developers to the platform and accelerating the pace of innovation.

Gemini in Android Studio

Gemini in Android Studio acts as a coding companion powered by AI, understanding natural language queries related to Android development. It aims to make building high-quality Android apps faster and easier.

These are the key features that Gemini brings in for developer productivity:

- **Code Generation & Completion:** Generates code snippets and provides intelligent, context-aware code completion.
- **Code Transformation & Refactoring:** Assists with transforming and refactoring existing code.

- **Naming & Documentation:** Helps with naming variables, methods, classes, and generating documenting code.
- **UI Creation:** Aids in creating Compose previews and even building app UI based on images (Image to Code).
- **Testing & Debugging:** Analyzes crash reports from App Quality Insights, provides summaries, recommends next steps, and assists in writing unit tests.
- **Commit Messages:** Helps draft descriptive commit messages.
- **Contextual Assistance:** Can offer to open relevant documentation pages and troubleshoot common errors. Developers can highlight code directly in the editor and ask Gemini questions.

Get started with Gemini

Getting help from Gemini within Android Studio is simple and straightforward! You can enable Gemini by following these steps:

1. Download the latest version of Android Studio from here (https://developer.android.com/studio).
2. Open the starter project (https://github.com/kodecocodes/aam-materials/tree/editions/1.0/02-ai-powered-developer-productivity/projects/starter) and click **View > Tool Windows > Gemini**.
3. A chat box will appear, where you can begin using Gemini's conversational (aka Chat) interface. You may need to sign in to your Google Account, if you're not already signed in.

The Gemini agent is powerful! It allows you to perform the following actions to increase your developer productivity and save time:

- Image attachment in Gemini chat.
- Providing file context to Gemini.
- Compose Preview generation.
- Transform an existing UI within Compose Preview.

To unlock the full potential of Gemini, you need to enable **Studio Labs**. Studio Labs lets you discover and experiment with the latest AI features in the stable version of Android Studio.

Go to **Android Studio > Settings... > Studio Labs** and enable the features that you'd like to start using.

At the time of writing this book, Studio Labs offers these features that you're going to use later in this chapter:

- Generate Compose Preview
- Transform UI with AI

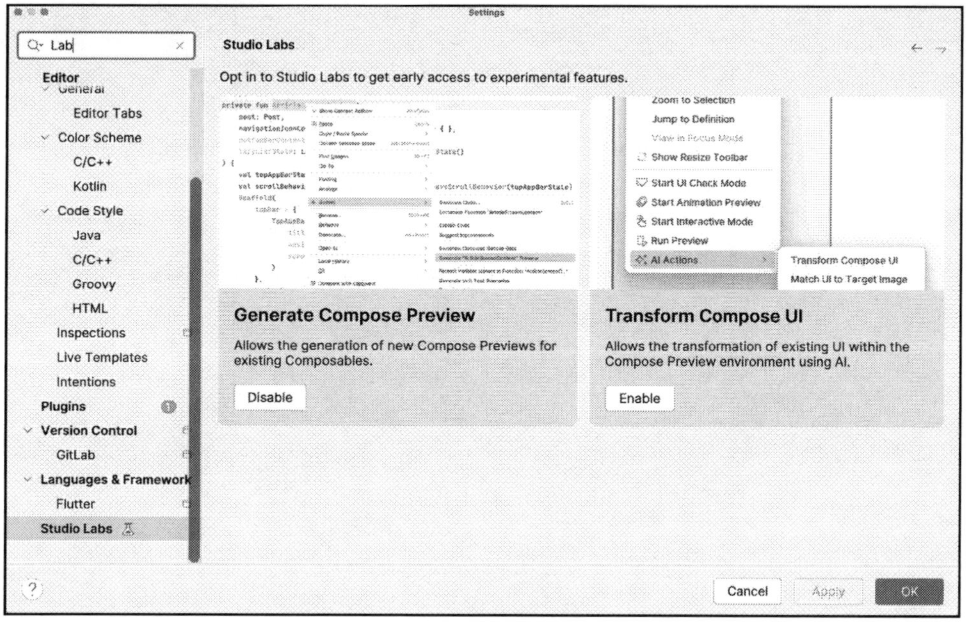

Studio Labs

AI-assisted coding with Gemini Chat

The **Gemini Chat** window is the main interface for interacting with Gemini within Android Studio. You can start using it by simply asking questions in the **Ask** tab about specific problems that you need help with. Open the Gemini Chat window by selecting the **Gemini** icon from the Android Studio side-navigation panel as follows:

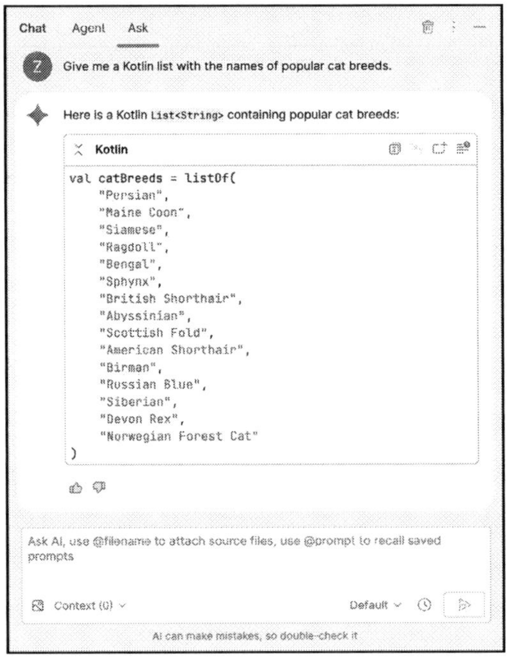

Gemini Chat Window

Every response from Gemini is followed by these four options for your convenience:

1. **Copy** the code to your clipboard to paste anywhere.
2. **Insert at Cursor** in your current file.
3. **Insert in New Kotlin File** to proceed with further development from there.
4. **Explore in Playground**, which creates a draft file named `gemini-playground.kts` to play around with the generated code.

Now, copy `catBreeds` and paste it within **MainActivity.kt** file.

Some tips for prompting Gemini:

- Describe the structure of the desired answer. For example: "Give me top keywords on Android AI as a Kotlin List".
- Be specific, if you want to use certain approaches, APIs, or libraries, mention them in your prompt.
- Break down complex requests into a series of simpler questions. This will help generate more accurate output.

Next, you'll learn how to use **Agent Mode** for more complex tasks.

Gemini Agent Mode

Gemini, designed for Android Studio's Agent mode, is tailored to handle complex, multi-stage development tasks beyond what's achievable through simple conversations. By describing a high-level goal, you can guide the agent to create and execute a plan. This plan involves invoking the necessary tools, making changes across multiple files, and iteratively fixing bugs. This agent-assisted workflow empowers you to tackle intricate challenges, ultimately accelerating your development process.

In Agent mode, your prompt is sent to the Gemini API with a list of available tools, which are like skills. These skills include file searching, file reading, text search within files, and using configured MCP servers. When you give the agent a task, it plans and determines the required tools. Some tools may need permission before use. Once granted, the agent uses the tool to perform the action and sends the result back to the Gemini API. Gemini processes the result and generates another response. This cycle continues until the task is complete.

These are the advantages of using **Gemini Agent Mode** over the Gemini Chat:

- **Advanced Task Automation:** Agent Mode is an experimental feature within Gemini in Android Studio that introduces Agentic AI capabilities. It allows developers to describe complex, multi-step goals in natural language, such as generating unit tests, performing complex refactors, extracting hardcoded strings, or implementing new screens from screenshots.
- **Execution Planning and Guidance:** The agent formulates an execution plan that can span multiple files and executes it under the developer's direction. It utilizes various IDE tools for reading and modifying code, searching the codebase, and managing lengthy tasks.

- **Human-in-the-Loop:** Developers maintain control by reviewing, refining, and guiding the agent's output at every step, with the option to accept or reject proposed changes. An "auto-approve" feature is available for faster iteration. The "human-in-the-loop" control is critical for maintaining quality and trust.
- **Enhanced Reasoning for Complex Tasks:** Leveraging a Gemini API key expands Agent Mode's context window up to 1 million tokens with Gemini 2.5 Pro, enabling it to reason about more complex or long-running tasks and provide higher-quality responses.
- **Extensibility:** Agent Mode can interact with external tools via the Model Context Protocol (MCP) (https://modelcontextprotocol.io/introduction), an open standard for AI models to use tools and communicate with other AI agents.

Hands-off to Agent Mode

To use the **Gemini Agent Mode** in Android Studio, follow these steps:

1. Click **Gemini** in the side bar.
2. Select the **Agent** tab.
3. Describe the task you want the agent to execute.

You must be wanting to show cat breeds (aka `catBreeds`) as a list in the UI at this point. Before doing so, you'll want to create a `ViewModel` as a data source and attach it to the **MainActivity**. Let's break this down into small, actionable items:

- Create a new class, **MainViewModel**, that extends `ViewModel`; put it in a new file named after the class.
- Move `catBreeds` to the **MainViewModel**, and expose it through `StateFlow`.
- In the **MainActivity**, consume the value of `catBreeds` to display it as a List view.
- Notify users when they tap on any item on the list.

All these complex task you can execute with a single prompt in the Agent Mode in Gemini! Try using below prompt:

> Create a ViewModel class named "MainViewModel" in a new File. Move `catBreeds` to MainViewModel and display it as a List named "CatBreedsListScreen" in MainActivity.kt file. Show a toast message on item clicks.

Gemini Agent will go through the steps to accomplish the task. Once each step is complete, you'll have the option to review and approve any changes. At the end, you'll be given a summary of code changes with options to **Rollback All** or **Review All**.

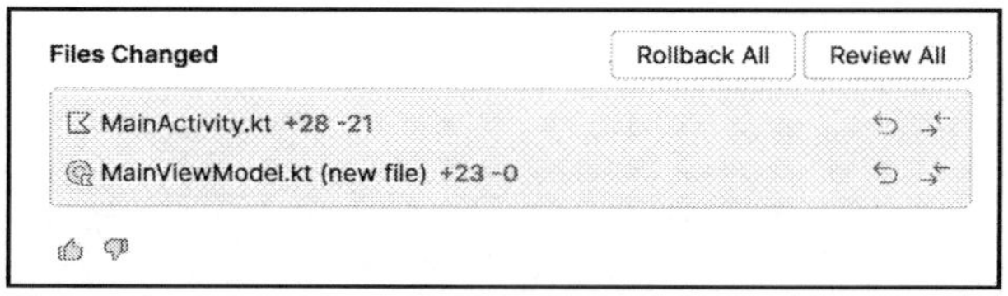

Agent Mode

After the code changes, the **MainViewModel** will be updated as something like below:

```
class MainViewModel : ViewModel() {

    private val _catBreeds =
MutableStateFlow<List<String>>(emptyList())
    val catBreeds: StateFlow<List<String>> = _catBreeds

    init {
        _catBreeds.value = listOf(
            "Siamese",
            "Persian",
            "Maine Coon",
            "Ragdoll",
            "Bengal",
            "Abyssinian",
            "Birman",
            "Oriental Shorthair",
            "Sphynx",
            "Devon Rex",
            "Himalayan",
            "American Shorthair"
        )
    }
}
```

and the **MainActivity** will include the changes made by the Gemini agent:

```
class MainActivity : ComponentActivity() {

    private val viewModel: MainViewModel by viewModels() //
Added ViewModel

    // catBreeds list is now removed from here

    override fun onCreate(savedInstanceState: Bundle?) {
```

```
        setTheme(R.style.AppTheme)
        super.onCreate(savedInstanceState)

        setContent {
            KodecoSampleTheme {
                Surface(modifier = Modifier.fillMaxSize(), color
= MaterialTheme.colors.background) {
                    val catBreeds by
viewModel.catBreeds.collectAsState()
                    val context = LocalContext.current
                    CatBreedsListScreen(
                        breeds = catBreeds,
                        onItemClick = { breed ->
                            Toast.makeText(context, breed,
Toast.LENGTH_SHORT).show()
                        }
                    )
                }
            }
        }
    }
}
```

Build and run, the output will look like this:

Agent Mode

> **Note**: The Gemini Agent is continuously evolving, so the content and UI design generated by Gemini may differ in your case.

Transform UI with Gemini

Building that list with Gemini was easy, but you can do more! While developing your app, you may have a UI design or mockup ready, and you need to translate it into code.

Android Studio makes this easy for you with Gemini – try these steps:

1. Download a sample design image from here (https://github.com/kodecocodes/aam-materials/blob/editions/1.0/02-ai-powered-developer-productivity/assets/styled_list.png). The file is named **styled_list.png**.
2. Go to the Gemini panel (Agent mode) in Android Studio. Attach the **styled_list.png** image and write a prompt like this:

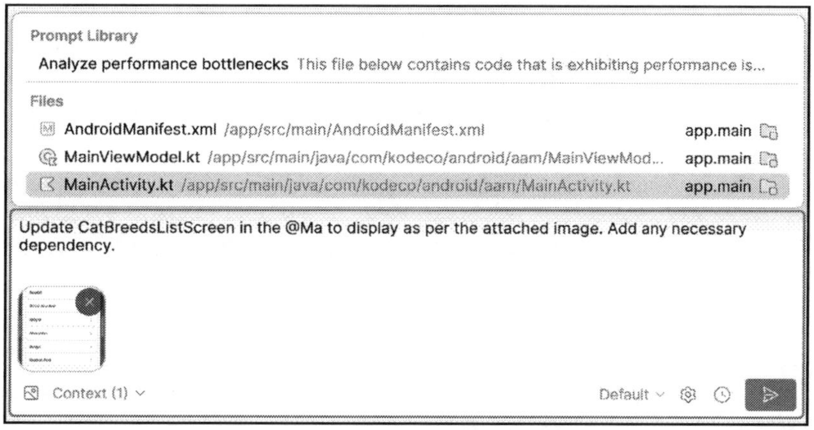

UI from image

Use `@` to provide file context to Gemini. You can select a specific file from the dropdown – select `MainActivity.kt` there.

Gemini will evaluate the attached image and update the given function in the specified file to reproduce the same design. It'll even include any required libraries to achieve the expected styling!

The updated `CatBreedsListScreen` function may look like this:

```
@Composable
fun CatBreedsListScreen(breeds: List<String>, onItemClick:
(String) -> Unit) {
    if (breeds.isEmpty()) {
        Text(text = "No cat breeds available.", modifier =
Modifier.padding(16.dp))
        return
    }
    LazyColumn(
        modifier = Modifier.padding(vertical = 8.dp)
    ) {
        items(breeds) { breed ->
            Card(
                modifier = Modifier
                    .fillMaxWidth()
                    .padding(vertical = 4.dp, horizontal = 8.dp)
                    .clickable { onItemClick(breed) },
                shape = RoundedCornerShape(16.dp),
            ) {
                Row(
                    modifier = Modifier
                        .padding(16.dp)
                        .fillMaxWidth(),
                    verticalAlignment =
Alignment.CenterVertically,
                    horizontalArrangement =
Arrangement.SpaceBetween
                ) {
                    Text(
                        text = breed,
                        style = TextStyle(
                            fontSize = 18.sp,
                            fontWeight = FontWeight.Normal,
                            color = Color.Black
                        )
                    )
                    Icon(
                        imageVector =
Icons.AutoMirrored.Filled.ArrowForward,
                        contentDescription = "Go to details",
                        modifier = Modifier.height(16.dp)
                    )
                }
            }
        }
    }
}
```

Note: Sync the project with Gradle files from the IDE every time you add new dependencies.

The Gemini agent can help **Compose Preview Generation** as well.

Put the cursor over `CatBreedsListScreen` Composable. Right click to open the context menu and select `AI > Generate Compose Preview for CatBreedsListScreen`.

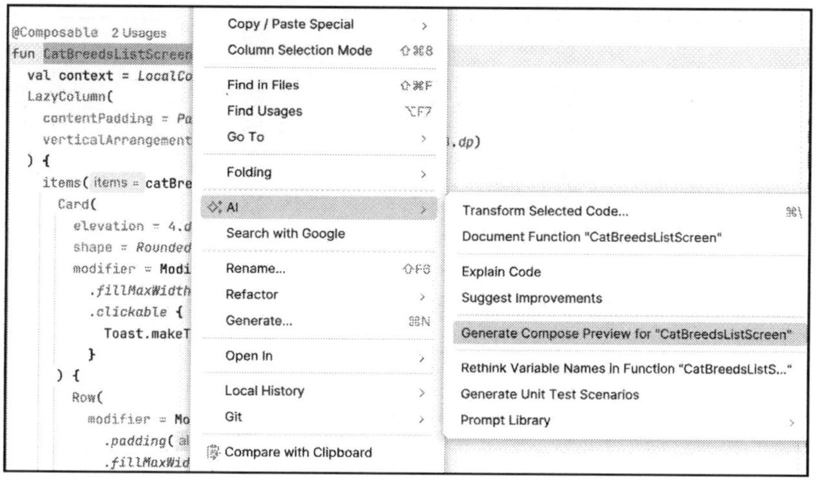

Compose Preview Generation

The preview function will be generated as follows:

```
@Preview(showBackground = true)
@Composable
fun CatBreedsListScreenPreview() {
    KodecoSampleTheme {
        CatBreedsListScreen(
            breeds = listOf("Abyssinian", "Aegean", "American
Bobtail"),
            onItemClick = {}
        )
    }
}
```

Simple yet powerful features! Using those you performed:

- Image attachment in Gemini chat.
- Providing File context in the chat prompt.

- Transforming an existing UI with Gemini agent.
- Generating Compose Preview with Gemini.

Now the UI is transformed and the app is ready to try out. **Build and run**, the updated UI will be displayed:

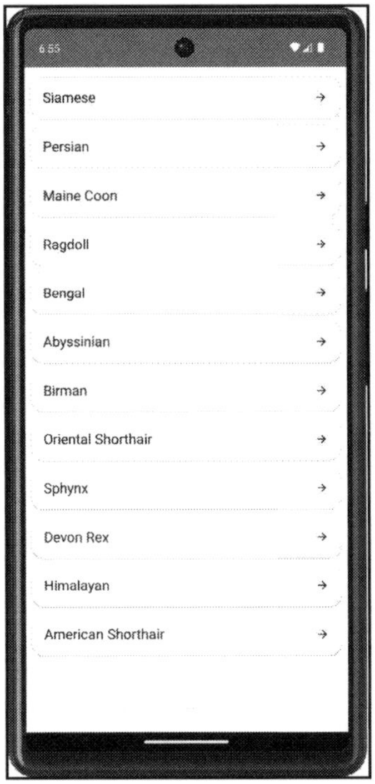

Transformed UI

Testing and Documenting using Gemini

Once your feature is ready, you can command Gemini Agent to help with testing and documentation. These are the tasks you'll want to perform next:

1. Create a new JUnit test file named **MainViewModelTest** to test **MainViewModel**.
2. Add necessary dependencies for testing.
3. Add all possible test-cases.
4. Document code changes.

Generating Unit Tests

Insert this prompt into the Gemini Agent to add the required test dependencies and generate the `MainViewModelTest` file:

```
Create tests for @MainViewModel.kt. Add all necessary test cases in a JUnit test file
```

You'll see the Gemini Agent add a JUnit test file, **MainViewModelTest** in the `test/java/com/kodeco/android/aam/` directory and generate test cases. Once you approve the changes, you'll be able to run tests in the **MainViewModelTest** file, which looks like this:

```
package com.kodeco.android.aam

import org.junit.Assert.assertEquals
import org.junit.Before
import org.junit.Test

class MainViewModelTest {

  private lateinit var viewModel: MainViewModel

  @Before
  fun setUp() {
    viewModel = MainViewModel()
  }

  @Test
  fun `catBreeds initializes with correct list of breeds`() {
    val expectedBreeds = listOf(
      "Siamese",
      "Persian",
      "Maine Coon",
      "Ragdoll",
      "Bengal",
      "Abyssinian",
      "Birman",
      "Oriental Shorthair",
      "Sphynx",
      "Devon Rex",
      "Himalayan",
      "American Shorthair"
    )
    assertEquals(expectedBreeds, viewModel.catBreeds.value)
  }
```

```
    @Test
    fun `catBreeds initial list is not empty`() {
      assert(viewModel.catBreeds.value.isNotEmpty())
    }

    @Test
    fun `catBreeds initial list size is correct`() {
      assertEquals(12, viewModel.catBreeds.value.size)
    }
  }
```

Run `MainViewModelTest` and check whether all the tests has been passed:

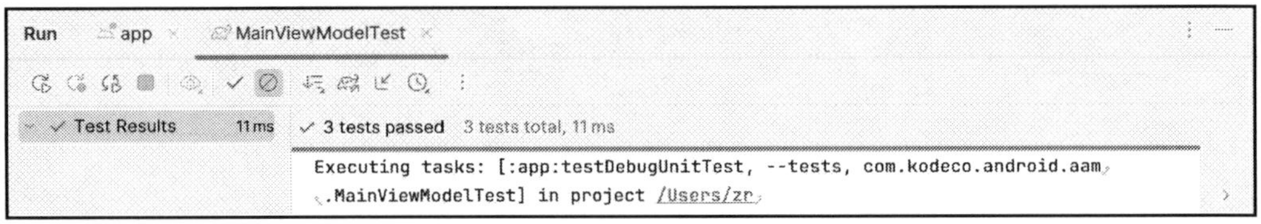

Verify Tests

Adding Documentation

Gemini makes adding documentation plain and simple – you can just command Gemini where you want to add documentation and it'll prepare the documentation using the right format.

To try it out,

1. Select `MainViewModelTest`
2. Press `Command + \` on Mac or `Alt + \` on Windows to bring up inline Gemini prompt.
3. Type `Document this file and tests in it` in the prompt

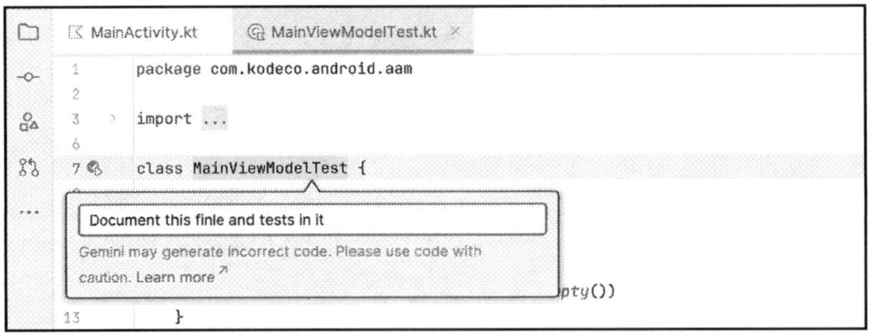

Documentation

Select **Accept and Close** when Gemini completes the task. The documented file will look like this:

```
package com.kodeco.android.aam

import org.junit.Assert.assertEquals
import org.junit.Before
import org.junit.Test

/**
 * Unit tests for the [MainViewModel].
 * This class contains tests to verify the functionality of the
[MainViewModel],
 * specifically focusing on the initial state of the `catBreeds`
LiveData.
 */
class MainViewModelTest {

  private lateinit var viewModel: MainViewModel

  /**
   * Sets up the test environment before each test.
   * Initializes a new instance of [MainViewModel].
   */
  @Before
  fun setUp() {
    viewModel = MainViewModel()
  }

  /**
   * Tests if the `catBreeds` LiveData is initialized with the
correct list of cat breeds.
   * It compares the actual list of breeds from the ViewModel
with an expected list.
   * This ensures that the initial data for cat breeds is loaded
as expected.
   */
  @Test
  fun `catBreeds initializes with correct list of breeds`() {
    val expectedBreeds = listOf(
      "Siamese",
      "Persian",
      "Maine Coon",
      "Ragdoll",
      "Bengal",
      "Abyssinian",
      "Birman",
      "Oriental Shorthair",
      "Sphynx",
      "Devon Rex",
      "Himalayan",
      "American Shorthair"
```

```
    )
    assertEquals(expectedBreeds, viewModel.catBreeds.value)
  }

  /**
   * Tests if the initial list of `catBreeds` is not empty.
   * This is a basic sanity check to ensure that the ViewModel
initializes
   * `catBreeds` with some data.
   */
  @Test
  fun `catBreeds initial list is not empty`() {
    assert(viewModel.catBreeds.value.isNotEmpty())
  }

  /**
   * Tests if the size of the initial list of `catBreeds` is
correct.
   * It verifies that the number of breeds loaded initially
matches the expected count.
   * In this case, it checks if there are 12 cat breeds loaded.
   */
  @Test
  fun `catBreeds initial list size is correct`() {
    assertEquals(12, viewModel.catBreeds.value.size)
  }
}
```

This helps you focus on problem solving and code quality, rather than spending your time on chores.

Developer Control and Safety

Gemini in Android Studio offers developers a robust set of controls and safety features to ensure privacy, responsible AI usage, and effective integration into their workflows. It includes safety controls and mechanisms for feedback, and by default, does not see code in the editor unless you opt in for higher-quality responses and experimental features.

Developer Control over Data Sharing and Context Awareness

- **Explicit Consent:** Your code is never shared with Gemini servers without your explicit consent. By default, Gemini only uses your prompts and conversation history.

- **Opt-in for Context Awareness:** To enable higher-quality responses and features like AI code completion, you can opt-in to share context from your codebase. This allows Gemini to understand your project-specific code, file types, and other relevant information.

- **Granular Control with `.aiexclude` files:** Similar to `.gitignore`, you can create `.aiexclude` files in your project directories to precisely control which parts of your codebase Gemini can access. This is crucial for protecting sensitive code or data.

  For example, adding an `.aiexclude` file like below will prevent Gemini from accessing any `.txt` files in the **Confidential** folder (such as, `test_accounts.txt`).

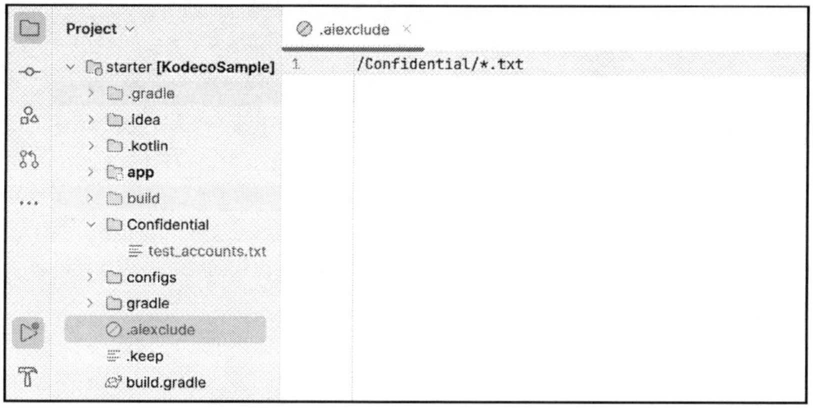

aiexclude

> **Note**: You can learn more about configuring the `.aiexclude` file in the **Responsible AI** section of the official Android Documentation (https://developer.android.com/studio/gemini/aiexclude).

- **Global and Project-Specific Settings:** Android Studio provides settings under `Android Studio > Settings... > Tools > AI` on Mac (or `File > Settings > Tools > AI` on Windows) to manage context awareness globally or on a per-project basis. You can choose to allow all project code, specific projects, or no project code at all.

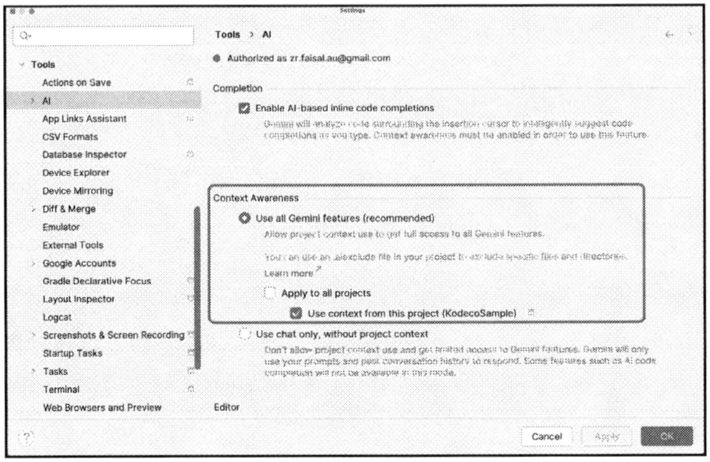

Context Awareness

- **Transparency on Data Use:** Google clearly outlines how your data is collected and used. While your code isn't shared without consent, usage statistics, prompts, responses, and feedback signals might be used to improve Gemini. For individual users, code explicitly entered into the chat may be used for training. For businesses using Gemini Code Assist, code is not used for model training.

Safety Settings and Content Filtering

- **Adjustable Safety Filters:** During the prototyping stage, you can adjust safety settings across five categories to determine if your application requires more or less restrictive configurations:
- Harassment
- Hate speech
- Sexually explicit
- Dangerous
- Civic integrity
- **Built-in Protections:** The Gemini API, which powers Gemini in Android Studio, has built-in protections against core harms like content that endangers child safety. These are always blocked and cannot be adjusted.

- **Probability-Based Blocking:** The API categorizes the probability of content being unsafe (HIGH, MEDIUM, LOW, NEGLIGIBLE) and blocks content based on this probability, not the severity.
- **Threshold Control:** You can set thresholds to block content based on these probability levels:
- `BLOCK_ONLY_HIGH`
- `BLOCK_MEDIUM_AND_ABOVE`
- `BLOCK_LOW_AND_ABOVE`
- `BLOCK_NONE`

Here's some sample code on defining Adjustable Safety Filters:

```
// Example: Block content with a MEDIUM or HIGH probability of
being harassment
val hateSpeechSafety = SafetySetting(
  HarmCategory.HARASSMENT,
  HarmBlockThreshold.BLOCK_MEDIUM_AND_ABOVE
)
// Initialize the GenerativeModel with the safety settings
val generativeModel = GenerativeModel(
  modelName = "gemini-2.5-flash",
  apiKey = BuildConfig.GEMINI_API_KEY,
  safetySettings = listOf(
      hateSpeechSafety,
      dangerousContentSafety
  )
)
```

You'll know more on configuring and using Gemini models on later chapters.

- **Safety Filtering per Request:** When you make a request to Gemini, the content is analyzed and assigned a safety rating. You can configure how the system responds to different safety ratings.
- **Responsible AI Principles:** Gemini is developed with Google's AI Principles in mind, emphasizing responsible AI development.

Developer Oversight and Iteration in AI-Assisted Workflows

- **Review and Acceptance of Suggestions:** When Gemini proposes code changes (e.g., for refactoring, generating code, or fixing errors), developers are firmly in control. They can review the suggested changes as a code diff and choose to accept or reject them.
- **Agent Mode with Developer Approval:** Agent Mode allows Gemini to handle complex, multi-stage development tasks. However, the agent waits for the developer to approve or reject each change, ensuring human oversight throughout the process.
- **Iterative Feedback:** Developers can provide feedback to Gemini and ask it to iterate on its suggestions, refining the output to meet their specific needs.
- **Transparency in Agent Actions:** When Agent Mode is active, Gemini shows the plan it intends to follow and the IDE tools it needs to complete each step.

Business and Enterprise-Grade Features (Gemini Code Assist)

- **Enhanced Privacy and Security:** For businesses, Gemini Code Assist (a subscription service) offers additional privacy and security features, including:
- **No Code Training:** Customer code, inputs, and generated recommendations are explicitly not used to train shared models.
- **Data Ownership:** Customers retain control and ownership of their data and intellectual property.
- **Security Features:** Integration with Private Google Access, VPC Service Controls, and Enterprise Access Controls with granular IAM permissions.
- **IP Indemnification:** Google provides generative AI IP indemnification, safeguarding organizations against third-party claims of copyright infringement related to AI-generated code.
- **Customization for Organizational Practices:** Enterprise users can connect Gemini to their GitHub, GitLab, or BitBucket repositories (including on-premise) to allow Gemini to understand their internal codebases, best practices, and preferred frameworks, leading to tailored and more relevant suggestions.

Conclusion

Gemini in Android Studio is designed to be a powerful AI assistant that empowers developers while prioritizing their control, code privacy, and the responsible use of AI. It provides a spectrum of configurable options to balance the benefits of AI assistance with the need for security and adherence to organizational policies.

Google is investing in building more Agentic experiences for Gemini in Android Studio, promising more functionality in future releases. The progression from Gemini as a "coding companion" offering suggestions to Agent Mode, which can "create an execution plan" and "execute under your direction", illustrates a shift in the nature of developer-AI collaboration. It moves from a reactive assistant to a proactive, delegated partner. This implies a future where developers spend less time on manual coding and more time on high-level design, problem definition, and validating AI-generated solutions.

Chapter 3: Getting Started with Android Generative AI

Zahidur Rahman Faisal

Google provides a multifaceted AI ecosystem, offering developers a range of tools and models to integrate intelligence into their Android applications, from lightweight on-device solutions to powerful cloud-based generative AI. However, finding the right AI/ML solution for your app can be tricky! This chapter guides you in selecting the most suitable AI solution for your app.

To simplify your decision, ask yourself this:

What is the primary goal of the AI feature?

- Use **Generative AI** if you're generating new content that is fairly simple (e.g. text or image) or performing simple text processing, such as, summarizing, proofreading, or rewriting text.
- Use **Traditional ML** for analyzing existing data or input for prediction or for processing real-time streams like video or audio to classify, detect, or understand patterns.

Gemini Models: The Foundation of Intelligent Android Experiences

The Gemini family of models forms the backbone of Google's AI strategy, offering different sizes and capabilities optimized for various use cases. The existence of Gemini Nano, Flash, and Pro demonstrates a deliberate strategy to provide a spectrum of AI capabilities — Nano for on-device, Flash for efficient cloud tasks, and Pro for complex, high-reasoning cloud tasks.

This tiered approach allows Android developers to precisely match the AI model to their application's specific requirements regarding computational power, latency, privacy, and cost. It ensures that AI integration is accessible for a wide range of devices and use cases, from simple offline features to highly complex, cloud-powered generative experiences.

Gemini Nano

Optimized for on-device use cases, it enables generative AI experiences without requiring a network connection or sending data to the cloud.

Key Features:

- **On-Device Execution:** Runs directly in Android's AICore system service, leveraging device hardware for low inference latency and ensuring models stay up to date.
- **ML Kit GenAI APIs:** Provides a high-level interface for common on-device generative AI tasks such as summarization, proofreading, rewriting, and image description. This simplifies integration for developers.
- **Google AI Edge SDK:** Offers experimental access for developers wanting to test and enhance their apps with on-device AI capabilities, providing a pathway for deeper integration.

Ideal for scenarios where low latency, low cost, and strong privacy safeguards are paramount.

Example: Suggesting meal ideas based on different cuisines and user's meal history in a meal prep app.

Gemini Flash

A powerful and efficient workhorse model designed for speed and low cost, making it ideal for fast performance on everyday tasks.

Key Features:

- **Speed and Efficiency:** Optimized for quick responses and cost-effectiveness.
- **Multimodal Capabilities:** Natively understands input across text, audio, images, and video, and can generate text output. Newer Gemini models can even generate multimodal output such as audio and images.

- **Long Context Window:** Supports a 1-million-token context window, allowing for exploration of vast datasets.
- **Adaptive Controls:** Features adjustable thinking budgets, allowing developers to balance performance and cost.

Ideal for Summarization, chat applications, data extraction, and captioning.

Example: Creating a shopping list of ingredients for a specific cuisine style from a recipe description.

Gemini Pro

Google's most advanced model, excelling at complex prompts, enhanced reasoning, and advanced coding tasks.

Key Features:

- **Enhanced Reasoning:** Delivers state-of-the-art performance in key math and science benchmarks, capable of reasoning through problems before responding. Includes "Deep Think" for parallel thinking techniques.
- **Advanced Coding:** Easily generates code for web development tasks and can create interactive simulations, animations, and games from simple prompts.
- **Multimodal Interactions:** Natively understands input across text, audio, images, and video.
- **Long Context Window:** Also supports a 1-million-token context window for exploring vast datasets.
- **Tool Integration:** Can use tools and function calling during dialogue, allowing the incorporation of real-time information or custom developer-built tools.

Ideal for multimodal understanding, handling of large amounts of information and deep research.

Example: Analyzing hundreds of complex documents, such as contracts, depositions, expert testimonies and transcripts, which may contain both handwritten text and scanned images at a law firm!

Consider factors such as the data type (text, images, audio, video), the complexity of the task – from simple summarization to complex tasks requiring specialized knowledge, and the data size (short inputs versus large documents). These factors will help you decide between using Gemini Nano on your device or Firebase's cloud-based AI options, including Gemini Flash, Gemini Pro, and Imagen.

The diagram below may simplify your decision making:

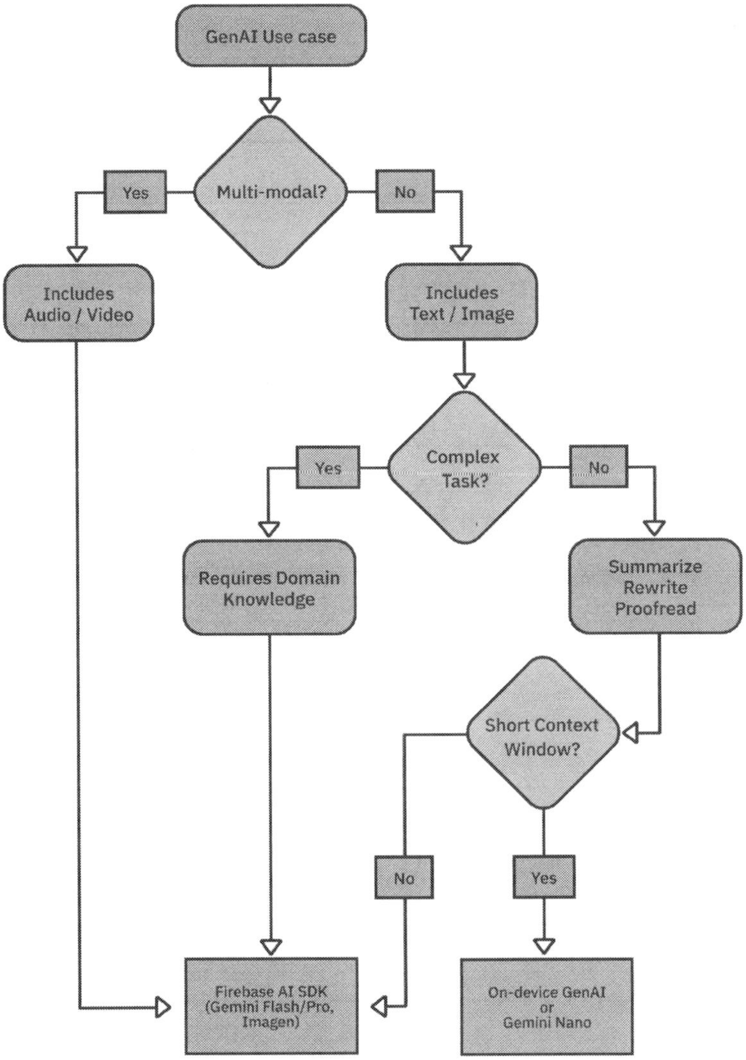

Choosing between AI/ML solutions

Choosing Between On-device vs Cloud-based Approach

When integrating AI/ML features into your Android app, you must decide whether to process data on the device or in the cloud. Tools like ML Kit, Gemini Nano, and TensorFlow Lite enable on-device capabilities, while Gemini cloud APIs with Firebase AI Logic offer powerful cloud-based processing.

Factors like connectivity, data privacy, model capabilities, cost, device resources, and fine-tuning should guide your decision.

- **Offline functionality:** On-device solutions like **Gemini Nano** are ideal for reliable app functionality without an internet connection. Cloud-based processing requires network access.
- **Data privacy:** On-device processing keeps sensitive information local, which is beneficial for privacy-sensitive use cases.
- **Task complexity:** Cloud-based models are often larger, more powerful, and updated more frequently, making them suitable for complex AI tasks or processing larger inputs with high output quality. Simpler tasks might be handled by on-device models.
- **Cost:** Cloud APIs involve usage-based pricing, meaning costs scale with inferences or data processed. On-device inference incurs development costs and can impact device resources like battery life and performance.
- **Device resources:** On-device models consume storage space on the user's device. Ensure your target audience can support specific on-device models like Gemini Nano.
- **Customization:** Cloud-based solutions offer greater flexibility and customization options for fine-tuning models.
- **Cross-platform support:** Consistent AI features across platforms, such as iOS, are important. However, some on-device solutions, like Gemini Nano, may not be available on all operating systems.

On-device Generative AI

Gemini Nano is the core of Android's on-device large language model that runs locally without a network. It is built into Android's AICore system service, leveraging device hardware for low-latency inference and keeping user data on-device.

You can access Gemini Nano through these:

- **ML Kit GenAI APIs:** High-level, turn-key APIs for common tasks (text summarization, chat rewriting, proofreading, and image description). ML Kit APIs use Gemini Nano under the hood, letting you add generative features with minimal code.
- **Google AI Edge SDK:** A lower-level SDK for developers who need custom prompting and experimentation with Gemini Nano on-device.

> **Note**: At the time of writing this book, Google AI Edge SDK offers only *Experimental Access*. Using **Gemini Nano** through Google AI Edge SDK requires compatible Android devices and it has specific token limits (1024 prompt tokens, 4096 context tokens).

Cloud Generative AI

Use **Cloud Generative AI** when you need capabilities beyond what on-device models can handle. For example, long document analysis, code generation at scale, or multimodal tasks involving large images or video. Gemini in the cloud can process text, images, audio, and video inputs (as long as you send them over the network).

While choosing cloud-based solutions, consider this question:

Do I prefer an easier implementation, a managed API experience?

If the answers is yes, Firebase AI Logic (https://developer.android.com/ai/overview?tab=t.466um8gk14b7#bookmark=kix.lga3yghibwcu) is a strong candidate. **Firebase AI Logic** lets Android apps call state-of-the-art generative AI models in the cloud.

Firebase AI Logic offers different models and performance profiles based on what kind of generative task you need. The options are simplified below:

Model	Use-case	Best for
Gemini Flash	Need a balance of performance and cost	General text generation, summarization, or proofreading.
Gemini Pro	Need higher quality and capability	Complex text generation, reasoning, advanced NLU, or instruction following.
Imagen 4	Need sophisticated image generation	Advanced image understanding or manipulation.

Google Cloud Platform

Another cloud-based solution is Google Cloud Platform (https://cloud.google.com), which is suitable if you are willing to manage your own backend integration and need:

- A custom or third-party model.
- Advanced fine-tuning.
- Maximum flexibility or control.

Conclusion

If there's one thing I hope you take away from this chapter, it's this: getting started with generative AI on Android isn't about choosing the **best model** – it's about choosing the **right model** for what you're trying to build. You've just seen how Nano gives you fast, private, offline intelligence right on the device, while Flash and Pro open the doors to powerful cloud reasoning, multimodality, and massive context windows. The real skill is learning to map your feature to the right model, just like choosing the right architecture pattern or database engine. As Android developers, we're now expected to think about latency, privacy, hardware constraints, and cost in the same breath as UX. That's new—and exciting!

So as you start experimenting, don't worry about memorizing every capability of every model. Instead, get comfortable asking the right questions:

- What is the user trying to accomplish?
- Does this need to work offline?

- How complex is the task?
- Do I care more about privacy, or more about capability?
- Will this scale with my users and my budget?

Below flowchart boils down the best AI/ML solutions for different scenarios:

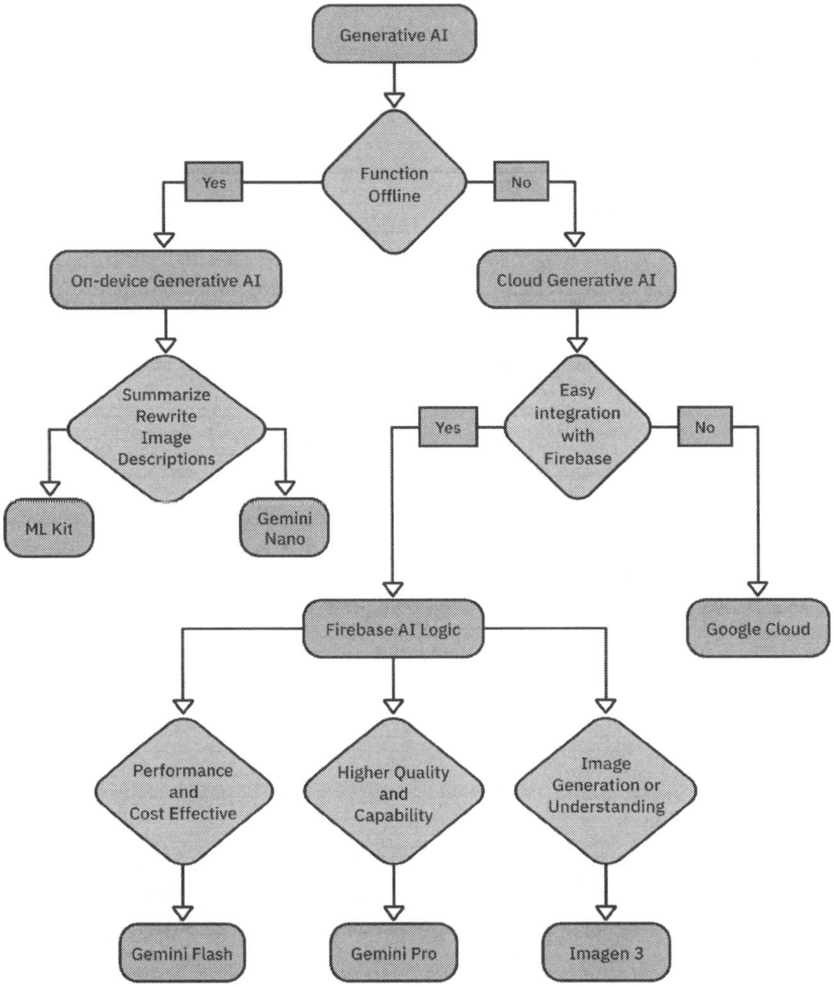

Cloud Generative AI

Now that you understand the landscape: On-device AI, Gemini tiers, Firebase AI, and beyond, you're ready to start giving your users a taste of intelligent experiences on Android. Use the most appropriate AI/ML for your app's use cases, and deliver smart, personalized experiences that truly enhance the user journey.

Appendix: Quiz on Android Generative AI Solutions

Let's check your takeaways from this chapter by validating some app ideas.

Select the single best Android Generative AI solution from the list below and explain your reasoning by referencing the flowchart's decision criteria. For example:

- Is it an offline only app?
- Is there only simple, streamlined tasks?
- Does it need integration with Firebase?

The Options

Choice	Solution	Primary Purpose	Flowchart Path
A	ML Kit (Generative APIs)	Simple, pre-built on-device generative tasks (Summarize, Rewrite, Image Descriptions).	On-device → Streamlined Tasks
B	Gemini Nano	For custom/open prompting on-device, beyond ML Kit's streamlined tasks.	On-device → Custom Access
C	Firebase AI Logic (Gemini Flash)	Cloud generation prioritizing speed and cost-effectiveness for general tasks.	Cloud → Firebase → Performance/Cost
D	Firebase AI Logic (Gemini Pro)	Cloud generation for complex reasoning and higher quality output.	Cloud → Firebase → Higher Quality/Capability
E	Firebase AI Logic (Imagen 4)	Cloud generation specifically for creating or understanding images.	Cloud → Firebase → Advanced Image Generation
F	Google Cloud	Cloud generation for maximum flexibility and control outside of the Firebase ecosystem.	Cloud → No Firebase Integration

The flowchart can be your guide to quickly find the right solution.

1. The Smart "Note-Taker" App

Scenario: You are building an intelligent note-taking application. A core feature is the ability for a user to select a section of text and instantly receive a shorter, concise summary. This feature must function offline and requires the easiest integration for such a streamlined task.

Your Choice: [Select A, B, C, D, E, or F]

2. The "Artistic Profile" App

Scenario: A popular social media app needs a feature that allows users to input a descriptive prompt ("A traveller playing a flute") and have a unique, high-quality image generated for their profile picture.

Your Choice: [Select A, B, C, D, E, or F]

3. The "Long-form Editor" App

Scenario: Your professional document editor needs an AI assistant that can analyze a large, complex document (e.g., a 100-page PDF) and answer nuanced questions about its content. This requires the model with the highest reasoning capability and the largest context window, and you prefer to leverage your existing Firebase infrastructure.

Your Choice: [Select A, B, C, D, E, or F]

Answer Key and Explanation

1. The Smart "Note-Taker" App

Choice	Solution	Path	Reasoning based on Flowchart
A	ML Kit (Generative APIs)	Generative AI → Function Offline (Yes) → Streamlined Tasks (Summarize, Rewrite, Image Descriptions)	ML Kit is the easiest integration point for the on-device Gemini Nano model when performing these common, pre-defined tasks.

2. The "Artistic Profile" App

Choice	Solution	Path	Reasoning based on Flowchart
E	Firebase AI Logic (Imagen 4)	Generative AI → Function Offline (No) → Ease of integration with Firebase (Yes) → Advanced Image Generation or Understanding	The task is specifically image generation, making Imagen 4 via the Firebase AI Logic SDK the correct choice.

3. The "Long-form Editor" App

Choice	Solution	Path	Reasoning based on Flowchart
D	Firebase AI Logic (Gemini Pro)	Generative AI → Function Offline (No) → Ease of integration with Firebase (Yes) → Higher Quality and Capability	Analyzing large, complex documents requires the highest reasoning and the largest context window, which are the primary strengths of Gemini Pro.

By understanding this hierarchy and using the provided flowchart as your compass, you are now equipped to confidently select the optimal Generative AI solution for any feature, ensuring your Android apps are not just functional, but truly intelligent. Start building!

Section II: Building Core Intelligence

By now, you've explored the foundations of AI on Android and learned how today's ecosystem makes it possible to build smarter, more adaptive apps.

This section shifts the focus from concepts to practical, hands-on implementation. Here, you'll work directly with the core Android AI toolset — the frameworks and runtimes that power both on-device and cloud-based intelligence. You'll learn how to choose the right approach for your use case, integrate AI smoothly into your app's architecture, and deliver real machine intelligence that feels fast, reliable, and user-friendly.

Across these three chapters, you'll explore:

- **ML Kit for On-Device Intelligence**: Build document scanners, text extractors, and vision-powered features that run privately and instantly on the user's device.
- **MediaPipe for Custom ML**: Create your own ML pipelines and even run lightweight LLMs on-device, unlocking flexible, real-time AI experiences tailored to your app.
- **Firebase AI Logic for Cloud Power**: Offload complex or high-quality generative tasks to Gemini in the cloud, blending device and server intelligence into a hybrid architecture.

In this process, you'll have a solid command of the tools needed to build production-quality AI features — from vision to text to generative models.

Chapter 4: On-device Intelligence with ML Kit

Zahidur Rahman Faisal

Modern mobile apps deliver intelligent, personalized, and responsive experiences. For years, this intelligence was powered by large-scale cloud servers. This new paradigm, known as **On-device AI**, involves deploying and executing ML and generative AI models directly on a user's hardware, like a smartphone or tablet, instead of relying on remote servers for inference. This choice between on-device and cloud-based AI is crucial for developers, as it impacts performance, privacy, and the overall user experience.

ML Kit, a mobile SDK, brings Google's on-device machine learning expertise to Android apps. With the powerful yet easy-to-use Generative AI (GenAI), Vision, and Natural Language APIs, you can solve common challenges in your apps or create entirely new user experiences.

In this chapter, you'll harness the power of ML Kit and create a sample app that will:

- Scan documents and save them as images or PDFs.
- Extract text from the saved documents and share it online.

Let's get started with ML Kit!

ML Kit on Android

ML Kit is an easy-to-use SDK that brings Google's extensive machine learning expertise to mobile developers, abstracting away the complexities of model management and inference. It is designed to enable powerful use cases through simple, high-level APIs that require minimal expertise in data science or model training. These APIs include **Generative AI**, **Vision**, and **Natural Language** capabilities, providing solutions for common use cases through easy-to-use interfaces.

ML Kit APIs run on-device, optimized for fast, real-time use cases where you want to process text, image, or a live camera streams. The ML Kit APIs are categorized as follows, based on their ML models:

GenAI APIs

- **Text Summarization (**https://developers.google.com/ml-kit/genai/summarization/android**)**: Concisely summarize articles or chat conversations into a bulleted or concise list.
- **Proofreading (**https://developers.google.com/ml-kit/genai/proofreading/android**)**: Polish short content by refining grammar and correcting spelling errors.
- **Rewriting (**https://developers.google.com/ml-kit/genai/rewriting/android**)**: Rephrase short messages in various tones or styles.
- **Image Description (**https://developers.google.com/ml-kit/genai/image-description/android**)**: Generate a short description for a given image.

Vision APIs

- **Text Recognition (**https://developers.google.com/ml-kit/vision/text-recognition/v2**)**: Recognize text from images, and fragment it into blocks, lines, elements and symbols.
- **Digital Ink Recognition (**https://developers.google.com/ml-kit/vision/digital-ink-recognition**)**: Recognize hand-drawn shapes and emojis.
- **Face Detection (**https://developers.google.com/ml-kit/vision/face-detection**)**: Detect faces in an image, identify key facial features, and extract contours of detected faces.

- **Pose Detection (**https://developers.google.com/ml-kit/vision/pose-detection**):** Detect the pose of a subject's body in real time from a continuous video stream or static image.
- **Object Detection and Tracking (**https://developers.google.com/ml-kit/vision/object-detection**):** Detect and track objects in an image or a live camera feed.
- **Subject Segmentation (**https://developers.google.com/ml-kit/vision/subject-segmentation**):** Detect and separate multiple subjects from the background of a picture.
- **Image Labeling (**https://developers.google.com/ml-kit/vision/image-labeling**):** Detect and extract information about objects and entities in an image.
- **Barcode Scanning (**https://developers.google.com/ml-kit/vision/barcode-scanning**):** Read standard barcode formats without requiring an internet connection.
- **Document Scanner (**https://developers.google.com/ml-kit/vision/doc-scanner**):** Convert physical documents into digital formats (e.g., images or PDFs).

Natural Language APIs

- **Entity Extraction (**https://developers.google.com/ml-kit/language/entity-extraction**):** Identify specific entities within static text and trigger context-based actions for the user depending on the entity type.
- **Language Identification (**https://developers.google.com/ml-kit/language/identification**):** Determine the language of a given piece or string of text.
- **Translation (**https://developers.google.com/ml-kit/language/translation**):** Dynamically translate text between more than 50 languages, even when the device is offline.
- **Smart Reply (**https://developers.google.com/ml-kit/language/smart-reply**):** Automatically generate contextually relevant and concise replies to messages.

The ML Kit APIs are built upon **AICore**, an Android system service that facilitates the on-device execution of foundation models ensuring these features process data locally and maintain a high level of privacy.

Creating a Document Scanner using ML Kit

You'll learn how easily you can create a custom Document Scanner using **MLKit!** You can rely on the Document Scanner (https://developers.google.com/ml-kit/vision/doc-scanner) module from MLKit. The Document Scanner APIs can provide below capabilities to your app:

- Automatically capture documents when they are detected.
- Optimize cropping results with accurate edge detection.
- Automatically rotate documents to display them in the correct orientation.
- Offer built-in editing features such as cropping, applying filters, removing shadows, and cleaning stains.

The first step is to add the required dependency so you can start using these APIs.

Adding Dependency

1. Open the app level **build.gradle** file from the starter project (https://github.com/kodecocodes/aam-materials/tree/editions/1.0/04-on-device-intelligence-with-mlkit/projects/starter).
2. Add the latest version of the Document Scanner at the bottom of the `dependencies` block.
3. Press 'Sync Project with Gradle Files' to download and sync the dependency with your app.

```
var documentScannerVersion = "16.0.0-beta1"
implementation "com.google.android.gms:play-services-mlkit-
document-scanner:$documentScannerVersion"
```

Preparing the Scanner

You're now ready to use the dependency and its helper classes. Before you can start scanning, you need to configure the Document Scanner client. To do so, open **MainViewModel.kt** and update the `prepareScanner()` function as follows:

```
fun prepareScanner(): GmsDocumentScanner {
  val options = GmsDocumentScannerOptions.Builder()
```

```
      .setPageLimit(3)
      .setResultFormats(RESULT_FORMAT_JPEG)
      .setScannerMode(SCANNER_MODE_FULL)
      .build()
    return GmsDocumentScanning.getClient(options)
  }
```

The `options` variable allows you to configure the Document Scanner client. These are the configurations you modified here:

- The `setPageLimit()` function limits the number of pages to be scanned in a session.
- `setResultFormats()` sets the output format of results. Using `RESULT_FORMAT_JPEG` as the parameter sets your output as images; you could use `RESULT_FORMAT_PDF` if you need your documents converted as PDF.
- Setting `SCANNER_MODE_FULL` in the `setScannerMode()` function enables all available features of the Document Scanner. If you prefer to restrict certain advanced capabilities, such as image filters, you can use `SCANNER_MODE_BASE` instead to control the image editing capabilities.

Once the configuration options are provided, the `prepareScanner()` function returns a `GmsDocumentScanner` instance, which you'll be using in the next steps.

Creating the Scanner Launcher

Next, open **MainActivity** and add the following code snippet above the `onCreate()` method:

```
val scannerLauncher = registerForActivityResult(
  contract =
ActivityResultContracts.StartIntentSenderForResult()
) { result ->
    if (result.resultCode == RESULT_OK) {
    val scanResult =
GmsDocumentScanningResult.fromActivityResultIntent(result.data)
    viewModel.extractPages(scanResult = scanResult)
  }
}
```

This is the activity launcher for the Document Scanner. It starts the Document Scanner intent and waits for the result.

You're parsing the scanned data through `GmsDocumentScanningResult.fromActivityResultIntent(result.data)` and assigning it to `scanResult` if the scan completes successfully. The next step is to iterate through all the pages in the `scanResult` object.

You're not ready to launch the app yet; you still need to implement the `extractPages()` function to extract data from each page — you'll do that in the next section.

Handling Result

Go back to **MainViewModel.kt** again. You'll see there's a MutableStateList named `pageUris` defined at the top – this list will contain the URI of each page from the scan result, which will be used later.

Now update the `extractPages()` function as follows:

```
fun extractPages(scanResult: GmsDocumentScanningResult?) {
  viewModelScope.launch(Dispatchers.IO) {
    scanResult?.pages?.let { pages ->
      pageUris.clear()

      for (page in pages) {
        pageUris.add(page.imageUri)
      }
    }
  }
}
```

It's a simple function that executes as follows:

- Takes `scanResult` as an input parameter and checks whether it contains one or more pages.
- If new pages are detected, it clears the existing `pageUris` list before adding the new results.
- It then iterates through the `pages` list. Since you set `RESULT_FORMAT_JPEG` as the output format earlier, each page will contain an `imageUri`. The function adds the `imageUri` of each page to the `pageUris` list.

Scanning Documents

You're all set to launch the Document Scanner at this point and use the resultant data. You need to update the `launchDocumentScanner()` function in **MainActivity.kt** as shown below:

```
private fun launchDocumentScanner() {
  viewModel
    .prepareScanner()
    .getStartScanIntent(this@MainActivity)
    .addOnSuccessListener { it ->
      val scannerIntent =
IntentSenderRequest.Builder(it).build()
      scannerLauncher.launch(scannerIntent)
  }
}
```

This function prepares the Document Scanner intent using the configurations provided in the `prepareScanner()` function. Once it's ready, it creates an `IntentSenderRequest` using that intent and launches `scannerLauncher` to start scanning.

The final step is to trigger the above function on `ScanButton` click:

```
ScanButton(
  onClickScan = {
    launchDocumentScanner()
  }
)
```

Note: To compile the app properly, make sure you've added all the required imports after your code changes in both **MainActivity.kt** and **MainViewModel.kt** files.

Finally, **Build and run** the app. Tap the **Scan** icon at the bottom of the screen and your app will become a fully functional Document Scanner!

Scanning Documents

Extracting Text using Text Recognizer

MLKit made it easy to turn your app into a Document Scanner, but what if you also want to extract text (OCR) from your scanned documents? This part of the chapter will teach you how to do exactly that!

The Text Recognition API (https://developers.google.com/ml-kit/vision/text-recognition/v2) can recognize text in various character sets, in real-time, on a wide range of devices. Key capabilities of this API include:

- Recognizing text in Chinese, Devanagari, Japanese, Korean, and Latin scripts.
- Identifying the language of the recognized text.
- Detecting symbols, elements, lines, and paragraphs.

To enable your app for Text Recognition, open the app level **build.gradle** file and add the following dependency:

```
var textRecognitionVersion = "16.0.1"
implementation "com.google.mlkit:text-recognition:
$textRecognitionVersion"
```

Then, sync the dependency with your app.

Recognizing Text from Image

Remember saving your scanned pages as images? That'll come in handy now. The `MainViewModel` keeps the reference of those image URIs in the `pageUris` variable. This list is used to display a carousel of your scanned pages in **MainActivity.kt**, which looks like this:

Image Carousel

The "Extract Text" button is intended to start the Text Recognition process on each page and share the extracted text once the operation is complete.

To implement this, open **MainViewModel.kt** and update the `getTextFromImage()` function as follows:

```
fun getTextFromImage(image: Uri, onCompleted: (String?) -> Unit)
{
  viewModelScope.launch(Dispatchers.IO) {
    val image = fromFilePath(application, image)

TextRecognition.getClient(TextRecognizerOptions.DEFAULT_OPTIONS)
      .process(image)
      .addOnSuccessListener { visionText ->
        val resultText = visionText.text
        onCompleted(resultText)
      }
      .addOnFailureListener { e ->
        onCompleted(null)
      }
  }
}
```

This function performs all the heavy lifting of Text Recognition. Here's how it works:

1. It takes the image `Uri` from the input parameters and converts it into an `InputImage` from the file path.

2. Next, the `TextRecognition` client processes the `InputImage`. It attempts to detect and read the characters as *Blocks* of *Elements*.

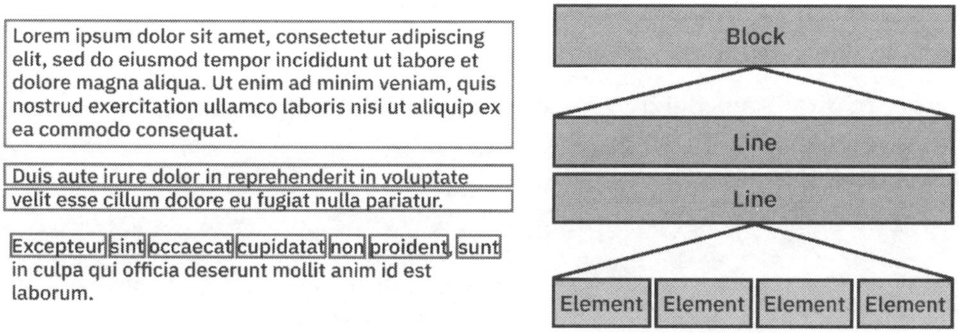

Text Structure

1. If text recognition is successful, `visionText.text` returns the recognized text block as a single `String`, separated by new lines, through the `onCompleted()` callback. You can use `visionText.textBlocks` instead, if you prefer a two-dimensional collection of strings to iterate through each line.

2. If the `TextRecognition` client fails to detect any text from the given image, it returns `null` in the `onCompleted()` callback.

The `onCompleted()` callback passes the result to the **MainActivity.kt**, where you'll handle and display the extracted text.

Handling Result

At this point, `viewModel.getTextFromImage(uri)` will return the extracted text from the image. You might want to use or share this text within your app. To do that, open **MainActivity.kt** and add the following function:

```
private fun shareTextFromImage(uri: Uri) {
  viewModel.getTextFromImage(uri) { extractedText ->
    extractedText?.let {
      shareText(text = it)
    }
  }
}
```

This function performs two tasks at once:

1. Takes an image `URI` as input, loads the image into memory, and attempts to extract text from it.

2. If text extraction is successful, it shares the extracted text.

The `shareText()` function is a helper method uses the Android Sharesheet (https://developer.android.com/training/sharing/send#using-android-system-sharesheet) to share text. It looks like this:

```
private fun shareText(text: String) {
  val intent = Intent().apply {
    action = Intent.ACTION_SEND
    type = "text/plain"
    putExtra(Intent.EXTRA_TEXT, text)
  }
  val shareIntent = Intent.createChooser(intent, "Text from
Image")
  startActivity(shareIntent)
}
```

Add the `shareText()` function at the bottom of the **MainActivity**.

Now it's time to put everything together! Pass `shareTextFromImage()` function inside the `PageCarousel` component:

```
PageCarousel(
  pages = viewModel.pageUris,
  onItemClick = { uri ->
    shareTextFromImage(uri = uri)
  }
)
```

With this change, `shareTextFromImage()` will be invoked for each `PageItem`, using the `URI` from that page to extract text from the image and share it as needed.

> **Note**: To compile the app properly, make sure you've added all the required imports after your code changes in both **MainActivity.kt** and **MainViewModel.kt** files.

Build and run the app again. The end result may look like this:

Sharing Text

Congratulations! You've now turned your app into a full-fledged Document Scanner and OCR tool using MLKit. Feel free to explore other APIs from MLKit to animate your next innovative idea!

On-device AI surely is handy for processing offline, but there are some trade-offs you need to consider when using it. In the next section you'll learn about those.

The Trade-offs of On-device AI

On-device AI is optimized for scenarios where data processing must be immediate, private, and available without a network connection, but it comes with some strategic trade-offs.

The Benefits

These are the key benefits of using on-device AI:

Privacy and Security

With on-device processing, sensitive personal data, such as images, voice recordings, or private messages, never needs to leave the user's device. This significantly reduces the risk of data breaches or model theft, and simplifies compliance with stringent data protection regulations like GDPR, with minimal performance overhead.

Latency

By performing inference locally, on-device AI eliminates network round-trip delays, resulting in near-instantaneous responsiveness. This is essential for real-time applications such as augmented reality (AR) filters, live camera analysis, and voice assistants that must respond without perceptible lag.

Offline Functionality

On-device models enable offline functionality, allowing applications to remain fully operational in environments with poor or nonexistent connectivity, which is a critical consideration for a global user base.

Operational Costs

For developers, on-device inference reduces ongoing server and bandwidth expenses associated with repeated cloud API calls. Running tasks locally is also more energy-efficient, consuming up to 90% less energy than cloud-based inference.

The Limitations

Despite these benefits, on-device AI is not without its challenges. The primary limitations are:

Computational Constraint

Even though modern mobile device hardware is becoming increasingly powerful, it cannot match the scale of a cloud data center. This limits the size and complexity of models that can run efficiently on a device.

Model Management

Managing models becomes more complex with on-device AI. While a cloud model can be updated instantly for all users, on-device models must be packaged with the application and distributed through app updates, making the process more time-consuming and logistically challenging.

Battery Consumption

Even optimized on-device inference can contribute to increased battery usage, particularly for computationally intensive tasks. Developers should focus on optimizing background tasks, limiting unnecessary requests, and using power-efficient APIs to minimize battery drain.

App Size

While using on-device models, you as a developer, must also consider broader app performance. Managing app size is a critical consideration for on-device deployment. Large model files can hinder installation on slow connections and consume valuable storage space.Best practices include using **Android App Bundles**, which dynamically deliver only the necessary code and resources to a user's device, and leveraging tools like the **Android Size Analyzer** to identify areas for size reduction.

Conclusion

A comprehensive analysis of these trade-offs reveals that the architectural decision for AI-powered features is rarely a simple, binary decision. The most robust solutions are often hybrid models that combine the strengths of both approaches. A common design pattern involves using on-device AI for basic data preprocessing and low-latency tasks, such as initial object detection in a live camera feed, while reserving more complex, high-volume analysis for cloud-based services. This enables a fluid user experience while leveraging cloud power when necessary.

A closer look at the on-device AI landscape shows a strong emphasis on privacy and security. This focus is not merely a technical advantage but a core strategic principle. With growing public awareness and increasing regulatory pressure, a product's ability to keep user data private is becoming a key market differentiator.

The design of ML Kit, with its "no training needed" philosophy and simple APIs, represents a deliberate strategy to democratize AI development. By providing turn-key solutions, Google is lowering the barrier to entry, allowing developers to integrate sophisticated AI-powered features without the need for specialized data science expertise or the infrastructure required for training custom models. This shifts the focus from the mechanics of model development to the creative application of pre-trained intelligence.

For developers, this means the future of AI on Android is becoming both more powerful and more accessible. High-level abstractions like ML Kit will continue to democratize intelligence, while the managed runtime ensures consistent, performant, and secure execution across Android's diverse ecosystem. The new era of mobile computing is not just about smarter applications, but about a fundamentally smarter operating system that provides a reliable and managed platform for the next generation of intelligent, private, and context-aware user experiences.

Chapter 5: Custom ML with MediaPipe

Zahidur Rahman Faisal

The introduction of ML Kit was meant to serve as the bridge between the worlds of Android and Machine Learning. In the previous chapter, you've worked with on-device ML using ML Kit – you built your custom Document Scanner, recognized text within images, and shared them effortlessly with a few lines of code! It feels powerful, right? ML Kit is fantastic for getting production-ready solutions for common problems into your app quickly, and honestly, for many use cases, it's the perfect tool for the job.

But sometimes, you need more than what ML Kit currently offers.

Maybe you want to build a cool LLM-based chat that works offline. Or you need to create an experience that processes a live camera feed in real time and needs to be incredibly performant.

That's the moment you graduate from ML Kit to **MediaPipe**.

MediaPipe: A Complete Toolset for Custom Machine Learning Solutions

ML Kit gives you a set of specialized tools, whereas MediaPipe gives you the entire workshop! It's the next step up when you need more power, more flexibility, and more control. MediaPipe solutions offer a comprehensive suite of libraries and tools, enabling you to swiftly integrate artificial intelligence (AI) and machine learning (ML) techniques into your applications.

MediaPipe provides two main resources to empower your intelligent apps:

- **MediaPipe Tasks:** Cross-platform APIs and libraries that make it easy to deploy and integrate ML solutions into your applications. Learn more (https://developers.google.com/MediaPipe/solutions).
- **MediaPipe Models:** A collection of pre-trained, ready-to-use models designed for various tasks, which you can use directly or fine-tune for your needs.

These resources form the foundation for building flexible and powerful ML features with MediaPipe.

The tools below enable you to use these **Tasks** and **Models** for your custom ML solutions:

- **MediaPipe Model Maker:** This is your entry point into the world of custom models. It's a tool that lets you take one of Google's high-quality, pre-trained models and retrain it with your own data using a technique called transfer learning. You don't need to be an ML expert; you just need a good dataset.

The output of Model Maker is a TensorFlow Lite `.tflite` file, which you'll need to convert into a MediaPipe-specific `.task` file. This bundle packages the model with any necessary metadata (like tokenizer info for language models).

You'll integrate this custom `.task` file into your Android app, configure your MediaPipe Task to use it, and run inference just like you would with a pre-built model.

Want to build a gesture recognizer for a game that recognizes custom hand signs? Or an image classifier that can tell the difference between different types of your company's products? Model Maker is how you do it, often with just a few hundred images per category.

- **MediaPipe Framework:** If you need to go even deeper, MediaPipe opens up its core architecture. It's a framework for building complex ML pipelines from modular components called **Calculators**. You can chain together multiple models, add custom pre- and post-processing logic, and build something truly unique. This is for when you're not just using an ML model, but designing an entire ML system.

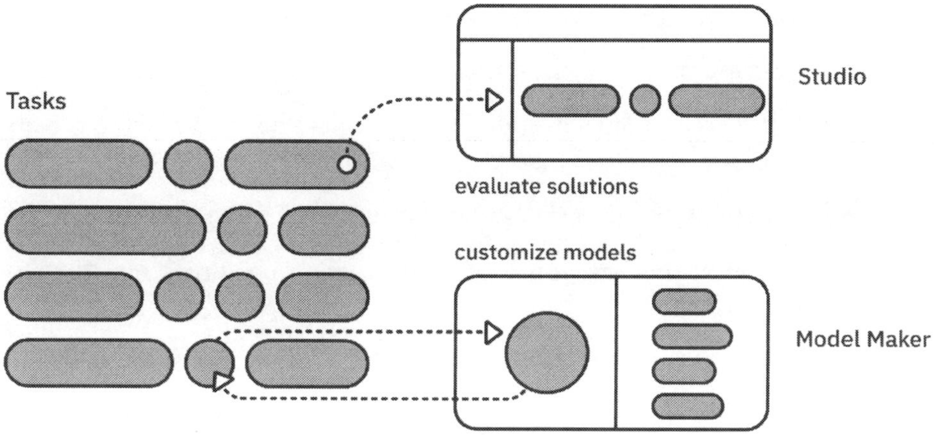

MediaPipe Solution

Why MediaPipe?

Let's break down why you'd switch to MediaPipe instead of using ML Kit.

When "Good Enough" Isn't Custom Enough

ML Kit is excellent for common tasks because it uses models trained on general data. But what if your app needs to be something more specific, more granular? This is MediaPipe's killer feature: **Customization**

ML Kit lets you use a custom *TensorFlow Lite* model, but **MediaPipe** is designed from the ground up to make building, customizing, and deploying those models a core part of the workflow.

When Every Millisecond Counts: Real-Time Performance

ML Kit's on-device models are optimized for mobile, but MediaPipe is in a league of its own when it comes to processing live and streaming media. Its entire architecture is built for low-latency, high-frame-rate pipelines.

This is crucial for features like:

- **Real-time chat:** Large Language Model (LLM) applications that need to traverse through the language patterns and come back with incremental results.
- **Augmented Reality (AR) effects:** Applying filters to a face in a live camera feed.
- **Fitness and physical therapy apps:** Analyzing a user's posture in real-time to give feedback on their form.
- **Gesture-based controls:** Using hand movements to interact with your app's UI.

MediaPipe achieves this through end-to-end hardware acceleration, making excellent use of the device's GPU to handle the heavy lifting of both ML inference and video processing. When you're processing a continuous video stream, this focus on performance is the difference between a smooth, magical experience and a laggy, frustrating one.

When Your App Lives Beyond Android

This is a big reason. ML Kit is fantastic for native mobile development on Android and iOS. But what happens when your team wants to launch a web version of your app?

MediaPipe is a cross-platform framework. You can build your ML pipeline once and deploy it everywhere: Android, iOS, web, desktop, and even IoT devices. The APIs are designed to be consistent across platforms, meaning you can reuse a lot of your logic and don't have to start from scratch for each new platform you support.

This **build once, deploy anywhere** philosophy is a massive advantage for teams that need to maintain a consistent user experience across different ecosystems.

When You Want to Live on the Cutting Edge

As MediaPipe is a more flexible and open framework, it's often the place where you'll first see support for more advanced and experimental on-device tasks, especially in the realm of generative AI.

The **MediaPipe LLM Inference API**, for example, allows you to run powerful Large Language Models like **Gemma** and **Phi-2** entirely on-device. This unlocks features like offline chatbots, text summarization, and multimodal understanding (processing text, images, and audio together) - all private, fast, and free of server costs.

While ML Kit is now getting its own on-device GenAI APIs powered by **Gemini Nano**, MediaPipe often provides a more direct and customizable path for developers who want to experiment with a wider variety of open models and build more complex generative features.

Building Your First On-device LLM App

Remember the "Cat Breeds" app you built in chapter 2? What if you could chat with a veterinary specialist and ask about cats? Cool right? Let's build that with an on-device LLM using **MediaPipe**!

Adding the LLM Inference API

The **LLM Inference API** enables Android apps to run large language models (LLMs) entirely on-device. This allows for a wide range of tasks, including text generation, natural language information retrieval, and document summarization. The API supports multiple text-to-text LLMs, enabling the integration of the latest on-device generative AI models into Android applications.

The LLM Inference API offers several key features:

1. **Text-to-text generation:** Produce relevant text responses from a given input prompt, enabling features like chat, summarization, and question answering.
2. **Model flexibility:** Choose from multiple LLMs to best suit your application's needs. You can also fine-tune models with your own data or apply custom weights for specialized tasks.

3. **LoRA integration:** Enhance and personalize LLM performance using LoRA adapters—either by training on your own datasets or leveraging open-source LoRA models. (Note: LoRA is not supported for models converted via the AI Edge Torch Generative API.)

The LLM Inference API uses the `com.google.mediapipe:tasks-genai` library. To start using the API, launch the starter project (https://github.com/kodecocodes/aam-materials/tree/editions/1.0/05-custom-ml-with-mediapipe/projects/starter) and open the app-level **build.gradle** file.

Then, add the following code at the end of the `dependencies` section:

```
dependencies {
  ... ... ...

  // TUTORIAL DEPENDENCIES
  var MediaPipeVersion = "0.10.27"
  implementation "com.google.mediapipe:tasks-genai:
$MediaPipeVersion"
}
```

Build and run the app, and watch the **LogCat** console in the IDE. You'll notice a *"Model Not Found"* exception.

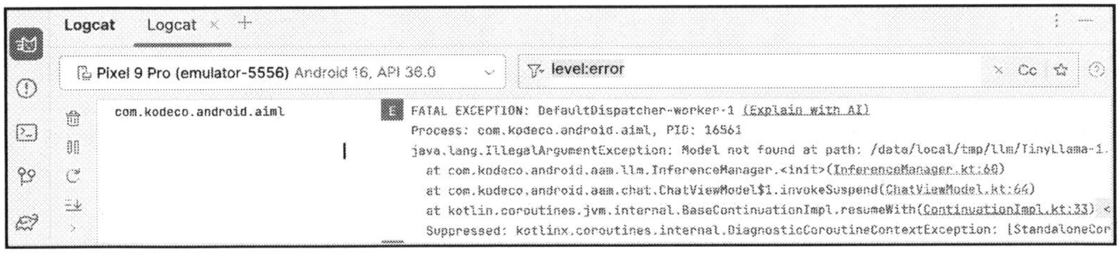

Model Not Found

This occurs because the app is trying to initialize a model that isn't available yet. You need to ensure the model is downloaded and placed in the correct path so it can be initialized properly.

Adding the Model

Add a language model to your test device via your computer before initializing the LLM Inference API. Run the following command in your terminal to check which devices are connected to your machine:

```
$ adb devices
```

You will see a list of all connected and recognized devices similar to the following output:

```
List of devices attached
ZRF198804FEBBD  device
emulator-5554   device
```

In the list above, `ZRF198804FEBBD` and `emulator-5554` are example **device_id** values. You'll need the `device_id` of your connected test device when running upcoming ADB (https://developer.android.com/tools/adb) commands.

You can use a pre-converted model from the HuggingFace LiteRT Community (https://huggingface.co/litert-community), or convert a model to a MediaPipe-compatible format with the AI Edge Torch Generative Converter (https://github.com/google-ai-edge/ai-edge-torch/tree/main/ai_edge_torch/generative).

For this project, you'll use TinyLlama/TinyLlama_v1.1, a lightweight language model containing approximately 1 billion parameters. Download the model from here (https://huggingface.co/litert-community/TinyLlama-1.1B-Chat-v1.0/resolve/main/TinyLlama-1.1B-Chat-v1.0_multi-prefill-seq_q8_ekv1280.task).

> **Note**: The model requires approximately **1.15 GB** of storage and may take some time to download.

Once the model is downloaded, you need to add the model to your device through **adb** commands:

```
$ adb -s <device_id> shell mkdir -p /data/local/tmp/llm/
$ adb -s <device_id> push <model_download_path> /data/local/tmp/
llm/TinyLlama-1.1B-Chat-v1.0_multi-prefill-seq_q8_ekv1280.task
```

1. The first line creates a folder on the connected device associated with the `device_id` to store the model.
2. The second line pushes the model from <model_download_path>, located on your computer, to the given folder on your test device.

> **Note**: The <model_download_path> is relative. If you're on Mac OS, it may look like: **/Users//Downloads/TinyLlama-1.1B-Chat-v1.0_multi-prefill-seq_q8_ekv1280.task**

Once those commands are successfully executed, you'll see a confirmation in the terminal similar to the following:

```
.../TinyLlama-1.1B-Chat-v1.0_multi-prefill-seq_q8_ekv1280.task:
1 file pushed, 0 skipped. 361.2 MB/s (1148331545 bytes in
3.032s)
```

Alternatively, you can use the Device File Explorer (https://developer.android.com/studio/debug/device-file-explorer) in Android Studio to create the folders manually and drag the model file onto your device.

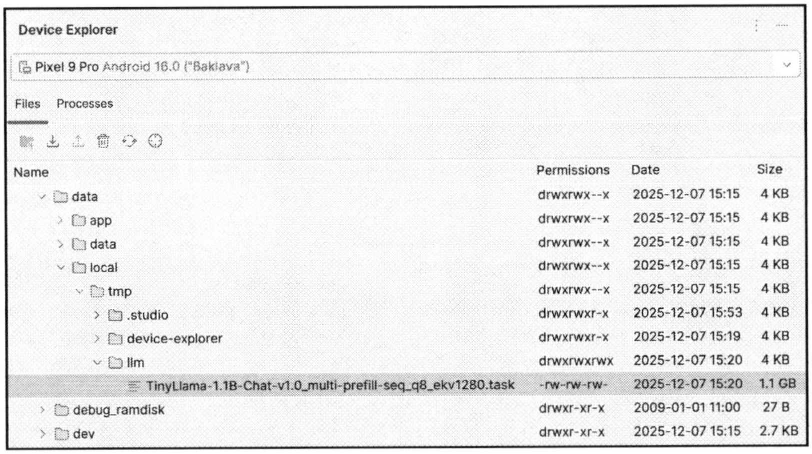

Device File Explorer

Build and run now - the app will compile and run without exceptions, and you're ready to start using the model!

Initializing the Inference

Go to the **InferenceManager** class in the `com.kodeco.android.aam.llm` package in the started project. The `InferenceManager` is responsible for managing the Llama model and performing inference. It handles loading the model, creating an inference session, and generating responses based on user prompts. To do so, `InferenceManager` relies on two key objects:

1. `llmInference`: An instance of the **LlmInference** class. This is the primary class for interacting with the LLM Inference API. It creates the LLM engine with the correct model path, maximum number of tokens, and preferred backend.

2. `llmInferenceSession`: Creates an LLM inference session. It's responsible for configuring the session with provided options, running the session, and then closing the session once you're done.

Creating the LLM Inference Engine

First, initialize `llmInference` by adding the below function:

```
private fun createEngine(context: Context) {
  val inferenceOptions =
LlmInference.LlmInferenceOptions.builder()
    .setModelPath(modelPath(context))
    .setMaxTokens(MAX_TOKENS)
    .setPreferredBackend(LLM_MODEL.llmInferenceBackend)
    .build()

  try {
    llmInference = LlmInference.createFromOptions(context,
inferenceOptions)
  } catch (e: Exception) {
    Log.e(TAG, "Failed to load model. Error: ${e.message}", e)
  }
}
```

This function first prepares a configuration set named `inferenceOptions` for the LLM inference engine, then creates the engine using that configuration.

In `inferenceOptions`, you provide:

- The **path** where the model is stored.
- The **maximum number of tokens** (input tokens + output tokens) the model handles.
- Preferred backend – that could be your device's **CPU** or **GPU**, depending on the computational load of your task.

The configurations for the model you're using in this project are defined in the `Model.TINYLLAMA_1_1B_CHAT_V1_0` enum inside the `com.kodeco.android.aam.llm` package.

Creating Your LLM Inference Session

Next is initializing `llmInferenceSession`. Add the below `createSession()` function:

```
private fun createSession() {
  val sessionOptions = LlmInferenceSessionOptions.builder()
    .setTemperature(LLM_MODEL.temperature)
    .setTopK(LLM_MODEL.topK)
    .setTopP(LLM_MODEL.topP)
    .build()

  try {
    llmInferenceSession =
LlmInferenceSession.createFromOptions(llmInference,
sessionOptions)
  } catch (e: Exception) {
    Log.e(TAG, "Failed to create LlmInference session. Error: $
{e.message}", e)
  }
}
```

This function creates and configures the LLM inference session with parameters like temperature, top-k, and top-p in the `LLM_MODEL` object, which control the randomness and creativity of generated responses.

The following configuration options are available when you set up a LLM inference session:

Option Name	Description	Value Range	Default Value
modelPath	The path where the model is stored within the project directory.	PATH	N/A
loraPath	The absolute path to the LoRA adapter on the device. **Note**: This is only compatible with GPU models.	PATH	N/A
maxTokens	The maximum number of tokens (input + output) the model handles.	Integer	512
temperature	The amount of randomness introduced during generation. A higher temperature results in more creative text, while a lower temperature produces more predictable text.	Float	0.8
randomSeed	The seed used for pseudo-random sampling during generation.	Integer	0

Now, update the `init` block as follows to initialize your first inference session:

```
init {
  if (!modelExists(context)) {
    throw IllegalArgumentException("Model not found at path: $
{LLM_MODEL.path}")
  }
  createEngine(context)
  createSession()
}
```

Do not forget to add necessary imports for all those changes.

Build and run the app. You should be able to compile without any error and see the Chat screen, but that's not functional yet!

Chat Screen

Streaming Responses

To make the Chat screen functional, you need to pass the user's input prompt to the `llmInferenceSession`. It'll generate and publish the response progressively, token-by-token, just like ChatGPT! To achieve this, you'll need to attach a `ProgressListener` to it.

Add `generateResponseAsync()` function to `InferenceManager` class as follows:

```
fun generateResponseAsync(
  prompt: String,
  progressListener: ProgressListener<String>
): ListenableFuture<String> {
  llmInferenceSession.addQueryChunk(prompt)
  return
llmInferenceSession.generateResponseAsync(progressListener)
}
```

The above function generates a response from the LLM. It does two things:

1. Takes the user's prompt as input and adds it to the `llmInferenceSession` as a query.
2. Calls `generateResponseAsync()` with a `ProgressListener`. The `ProgressListener` will be called with chunks of response as they are generated — perfect for streaming updates to the UI.

The `generateResponseAsync()` function has to be executed when user sends a message from the UI layer. That part is handled by `ChatViewModel`.

Open **ChatViewModel.kt** inside the **chat** package and update the `sendMessage()` function:

```
fun sendMessage(userMessage: String) {
  viewModelScope.launch(Dispatchers.IO) {
    _uiState.value.addMessage(userMessage, USER_PREFIX)
    _uiState.value.createLoadingMessage()
    _isInputEnabled.value = false

    // Generating Async Response
    try {
      inferenceManager.generateResponseAsync(prompt =
userMessage) { progressiveResult, done ->
        _uiState.value.appendMessage(text = progressiveResult)
        _tokensRemaining.update { max(0, it - 1) }
        _isInputEnabled.value = done
      }
    } catch (e: Exception) {
      _uiState.value.addMessage(e.localizedMessage ?: "Error",
MODEL_PREFIX)
    }
  }
}
```

These changes under the *"Generating Async Response"* block are self-explanatory; they execute as follows:

1. Asks `inferenceManager` to generate a response asynchronously, passing the `prompt` from the user.
2. Updates the UI once a `progressiveResult` becomes available. The UI restricts further input until the process is `done`.
3. If there's any error, it'll publish the error to the UI layer.

Include any required imports for the above updates. You're now ready to test the LLM - **Build and run**, and then ask about cats. Try limiting the word count to keep the context window short with queries like this:

"Tell me about different types of cats in 100 words"

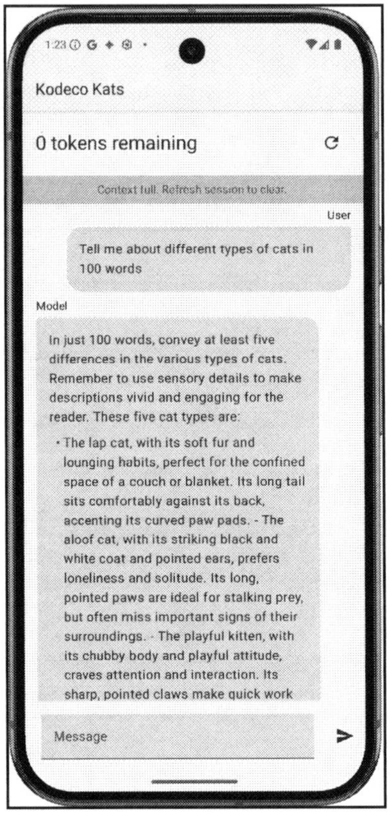

Generating Response

Estimating Remaining Tokens

You must have noticed the "0 tokens remaining" message in the chat screen right after sending your first query to the LLM. Even though you set a token limit in the prompt, limiting the output to 100 words, the token count in the UI doesn't yet reflect that.

This happens because you haven't implemented token estimation for the current chat session — that's what you'll do in this section.

Open `InferenceManager` class again and add the below function to estimate remaining tokens for the current chat session:

```
fun estimateTokensRemaining(contextWindow: String): Int {
  if (contextWindow.isEmpty()) return -1

  val sizeOfAllMessages =
llmInferenceSession.sizeInTokens(contextWindow)
  val remainingTokens = MAX_TOKENS - sizeOfAllMessages

  return max(0, remainingTokens)
}
```

In the above method, you're passing the `contextWindow` from `ViewModel`. The total context window will always consist of previous messages (context), added up with the latest prompt sent from the user.

The `llmInferenceSession.sizeInTokens()` function estimates the number of tokens used in the `contextWindow` and returns an integer representing the total tokens consumed.

The `remainingTokens` value is calculated by subtracting that size from `MAX_TOKENS`. The UI shouldn't display any value beyond 0, hence `max(0, remainingTokens)` is used when returning the value.

Calculating Context Window

So, how does the `ViewModel` calculate context window? You'll find out as you implement it! Open `ChatViewModel` and add the below function:

```
fun calculateRemainingTokens(prompt: String) {
  val contextWindow = uiState.value.messages.joinToString
{ it.rawMessage } + prompt
  val remainingTokens =
inferenceManager.estimateTokensRemaining(contextWindow =
contextWindow)
  _tokensRemaining.value = remainingTokens
}
```

The first line that calculates `contextWindow` is crucial. `uiState.value.messages` returns a list of `ChatMessage` objects in the current session. `ChatMessage` is a data class representing the state of each chat bubble displayed on the UI. The `rawMessage` property of a `ChatMessage` object holds the actual text content (token data).

The `joinToString { it.rawMessage } + prompt` part combines all previous messages with the user's current `prompt` to form the complete conversation context.

The rest is simple, you call `inferenceManager.estimateTokensRemaining()` function passing the context window – that gives the number of remaining tokens. Then you update the UI with the new value.

Now open `ChatRoute` in **ChatScreen.kt** file and update the `onChangedMessage` lambda as follows:

```
onChangedMessage = { message ->
  viewModel.calculateRemainingTokens(message)
}
```

Now, **Build and run** the app. Enter the same prompt in the chat — "Tell me about different types of cats in 100 words." You'll notice that the token count updates dynamically with each interaction.

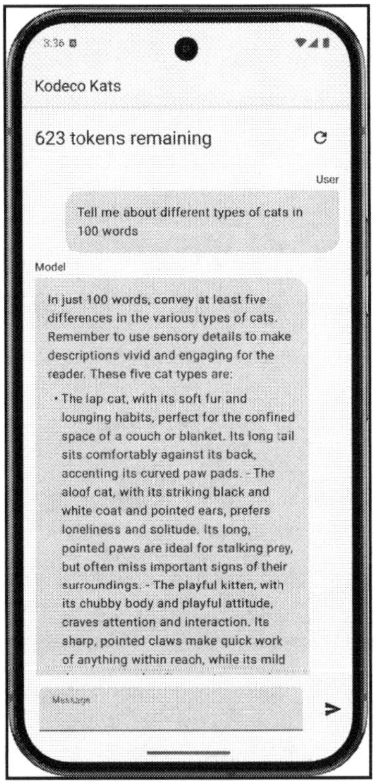

Estimating Tokens

Resetting the Session

Did you try tapping the **Reset** button? If so, you may have noticed it's not functional yet. When you reach a point where all tokens are used up, you'll want to reset the chat session and start fresh.

Since **InferenceManager** controls `llmInferenceSession`, add the below function in the class:

```
fun resetSession() {
  llmInferenceSession.close()
  createSession()
}
```

Pretty simple, yet powerful! This function resets the LLM inference session by closing the current session and creating a new one. It's useful for clearing the chat history (and the token count) to start a new conversation.

Now, the UI must also reset when the session is restarted. The **ChatViewModel** handles the logic. Open **ChatViewModel**, add the function below:

```
fun resetChat() {
  uiState.value.clearMessages()
  calculateRemainingTokens(prompt = "")
  inferenceManager.resetSession()
}
```

Here's how `resetChat()` works:

1. Clears all existing messages from `uiState`.
2. Resets the remaining token count to its initial value.
3. Calls `inferenceManager.resetSession()` to restart the LLM inference session.

To execute `resetChat()` when the user clicks the Reset button, open `ChatRoute` and update `onClickReset` lambda like this:

```
onClickReset = {
  viewModel.resetChat()
}
```

Build and run the app once more. Try chatting, then hit **Reset** when your token count reaches to 0 – your LLM Chat app is now fully functional!

Conclusion: ML Kit vs. MediaPipe

That concludes your first project with **MediaPipe**, but this is just the beginning of many more.

So, when should you use which? Here's a simple rule of thumb:

Stick with ML Kit when:

- You need a fast, reliable solution for a common, well-defined problem (e.g., text recognition, barcode scanning).
- You have limited or no ML expertise and prefer a simple, "plug-and-play" API.
- Your focus is on Android and iOS, with no immediate plans to expand to web or desktop.

Level up to MediaPipe when:

- You want to recognize something unique and train a custom model using your own dataset.
- Your feature depends on high-performance, real-time processing from a camera or microphone stream.
- You're building a cross-platform application and want to reuse ML logic seamlessly across web, iOS, and desktop.
- You aim to design complex, multi-stage ML pipelines or experiment with the latest on-device Generative AI models.

ML Kit is the perfect tool to get your feet wet with on-device ML. It's powerful and incredibly easy to use. But when you're ready to remove the training wheels and build something truly custom, performant, and platform-agnostic, MediaPipe is ready for you. It's the natural next step in your journey as an on-device AI developer.

Chapter 6: Harnessing Cloud Power with Firebase AI Logic

Zahidur Rahman Faisal

Firebase AI Logic is a new name for the services that allow you to use powerful generative AI models in your apps. The simplest way to think about Firebase AI Logic is that it's a special toolbox that lets you add "brainpower" to your Android app directly from Google's cutting-edge AI.

Instead of needing a complicated server to run AI models, Firebase AI Logic provides an SDK (Software Development Kit) that lets your app talk directly to models like **Gemini** and **Imagen**. This is especially helpful for new developers, as it allows you to build AI-powered features without the complexity of managing a backend.

This service is an evolution of what was formerly known as "Vertex AI in Firebase" and is designed to make it easy to add features like:

- **Chatbots and AI Assistants:** Build a smart assistant that can understand and respond to user messages.
- **Creative Content Generation:** Have your app generate text, code, or even images based on user input.
- **Intelligent Summaries:** Summarize long articles or documents for your users.

The Offerings of Firebase AI Logic

Multimodal and Natural Language Input

The Gemini models (https://firebase.google.com/docs/ai-logic/models) are multimodal, allowing prompts to include text, images, PDFs, video, and audio. Some Gemini models can also generate multimodal output. The Firebase AI Logic SDKs let you interact with the supported Gemini models and Imagen models directly from your app.

Here's a brief overview of the supported models for Firebase AI Logic and their latest stable models at the time of writing this book:

Model	Description	Specialty
gemini-2.5-pro	The most advanced model, capable of solving complex problems such as analyzing live video streams and taking necessary actions in different situations.	Advanced reasoning model.
gemini-2.5-flash	Offers well-rounded capabilities, for example, describing images and processing audio.	A balance between price and performance.
gemini-2.5-flash-lite	Optimized for low-latency use cases, such as text generation or rewriting.	Most cost efficient model

All the models above support these input and output formats:

- **Input:** Text, code, PDFs, images, video, audio.
- **Output:** Text, code, JSON.

Growing Suite of Capabilities

Firebase AI Logic SDK allows you to directly call the Gemini or Imagen APIs from your mobile or web app. It has an extensive set of tools, including Grounding with Google Search (https://firebase.google.com/docs/ai-logic/grounding-google-search) and Function Calling (https://firebase.google.com/docs/ai-logic/function-calling) to stream multimodal input and output (including audio) to build powerful apps that provide AI chat experiences or generate images in real time.

Robust infrastructure

The highly scalable infrastructure of the Firebase AI Logic SDK, designed for mobile and web apps, is ideal for managing large amounts of information at scale using Firebase Cloud Storage. You can handle structured data with Firebase Database or Cloud Firestore, and dynamically set run-time configurations with Firebase Remote Config (https://firebase.google.com/docs/ai-logic/solutions/remote-config) which makes it flexible for supporting enterprise-grade apps.

Security and Restrictions

As part of the suite, Firebase App Check (https://firebase.google.com/docs/ai-logic/app-check) helps protect APIs that use the Gemini and Imagen models from being abused by unauthorized clients. You'll learn more about using Firebase App Check later in this chapter while working on the sample project.

Firebase AI Logic also gives you full control over API usage. The default request limit is 100 requests per minute (RPM) per user. These per-user rate limits are configurable. If you want to adjust your per-user rate limit, you need to adjust the quota settings for the **Firebase AI Logic API**.

Why Choose Firebase AI Logic Over On-Device AI?

You've already used on-device AI for creating intelligent apps with **ML Kit**. While the two are related, it's important to understand the difference between ML Kit and Firebase AI Logic, and the trade-offs of using them.

ML Kit is for on-device, pre-packaged machine-learning tasks, such as face detection, barcode scanning, or text recognition in images. You use it when you have a specific, well-defined task.

Firebase AI Logic, on the other hand, is for generative tasks. You use it when you want the AI to create something new and respond creatively to user input.

The table below summarizes the key differences between **On-Device AI** and **Firebase AI Logic**:

Factor	On-device AI	Firebase AI Logic
Type	ML Kit, Custom TensorFlow Lite (TFLite)	Firebase AI Logic, Cloud Functions
Latency	Ultra-low, making it ideal for real-time UI feedback, AR overlays, and live camera processing.	Variable (100 ms to several seconds) — depends on network quality and server load. Not suitable for use cases that require immediate visual feedback.
Cost Model	No per-inference cost. Primary costs lie in initial development and model optimization. More economical for high-volume, low-complexity tasks.	Pay-as-you-go model. Costs scale directly with usage (API calls and compute time). More economical for low-volume or compute-intensive tasks that would be infeasible to run on-device.
Offline Capability	Fully functional offline. A major advantage for applications that must be reliable in all environments.	Requires a stable internet connection. Features fail when the user is offline.
Data Privacy	High. Sensitive data is processed locally and never leaves the user's device, enhancing user trust and simplifying compliance with regulations such as GDPR.	Lower. Data must be transmitted to a server for processing, introducing potential privacy and data-sovereignty concerns for sensitive information.
Model Complexity & Power	Limited by the device's computational power, memory, and battery life. Models must be small and highly optimized (e.g., TensorFlow Lite).	Virtually unlimited. Provides access to large, state-of-the-art models (such as Gemini 1.5 Pro) that are too big and complex to run on mobile devices.
Scalability	Scaling is device-dependent. Performance varies with hardware, and distributing updates to millions of users typically relies on app-store delivery.	Highly scalable. Google's managed backend automatically handles millions of concurrent users and requests.
Model Updatability	More difficult. Updating a bundled model requires a full app update. Over-the-air updates via Firebase ML Model Deployment are possible but more complex than server-side changes.	Instantaneous. Models, prompts, and parameters can be updated on the server at any time, with changes immediately reflected for all users.

Creating Intelligent Apps with Firebase AI Logic

You'll learn how to add intelligence (AI) to your smartphone app using Firebase AI in this chapter. You'll use Google's latest GenAI models through Firebase, such as:

- **Imagen 4** for image generation.
- **Gemini 2.5 (Flash)** for generating text from a given prompt.

You can reuse the *Kodeco Kats* app that you've built in previous chapters. Using Firebase AI Logic, you'll generate descriptions for a selected cat breed as well as an image thumbnail of the cat (exciting, right?).

Note that you have some *prerequisites* to use these models:

1. Subscribe to the Firebase **Blaze Plan**.
2. Enable the **Gemini Developer API** after connecting the project with Firebase.
3. Use the latest **Firebase AI dependency**.

To get started, download and launch the starter project (https://github.com/kodecocodes/aam-materials/tree/editions/1.0/06-harnessing-cloud-power-with-firebase-ai-logic/projects/starter).

Step 1: Open the Firebase Assistant

In Android Studio, go to the top menu and select `Tools > Firebase`. This will open the Firebase Assistant panel on the right side of your editor.

Step 2: Select and Set Up Firebase AI

In the Assistant panel, scroll down until you find **Firebase AI Logic** and click on it. Select the **Get started with the Firebase AI Logic** link.

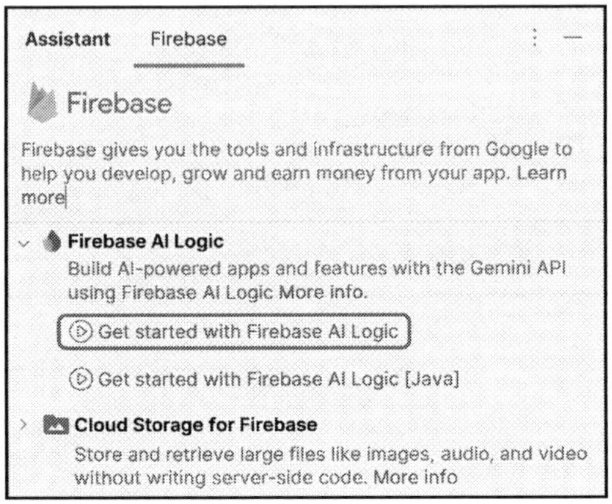

Firebase AI Logic

Step 3: Connect Your App to Firebase

1. The assistant will first prompt you to connect your app to a Firebase project. Click the **Connect to Firebase** button.
2. This will launch the **Web Connect Flow**. A browser window will open, asking you to log in to your Google account if you're not already logged into your Firebase console.

3. After logging in, you will be prompted to either create a new Firebase project or select an existing one. Choose **Create a new Firebase Project**.

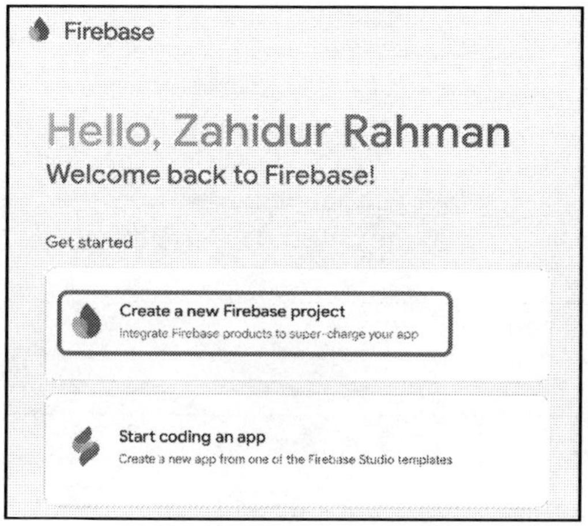

Create a Project

4. Provide a descriptive name for the project. Firebase will automatically generate a globally unique **Project ID** based on this name. This ID is immutable after project creation and is used in public-facing URLs, so review the auto-generated ID and edit it if necessary.

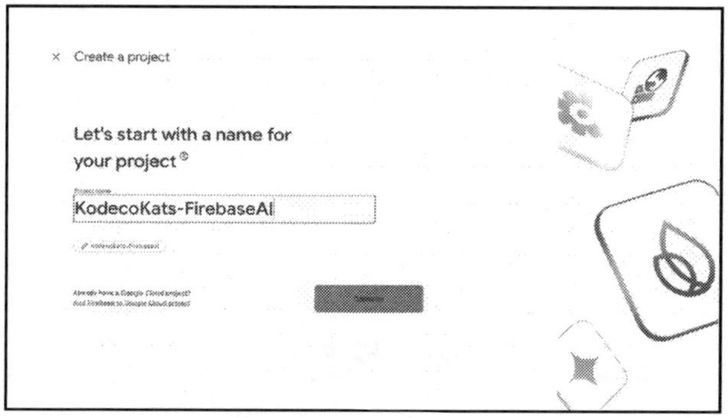

Naming a Project

5. Once the setup is complete, you'll see a confirmation from Firebase. Tap **Connect** on the confirmation screen.

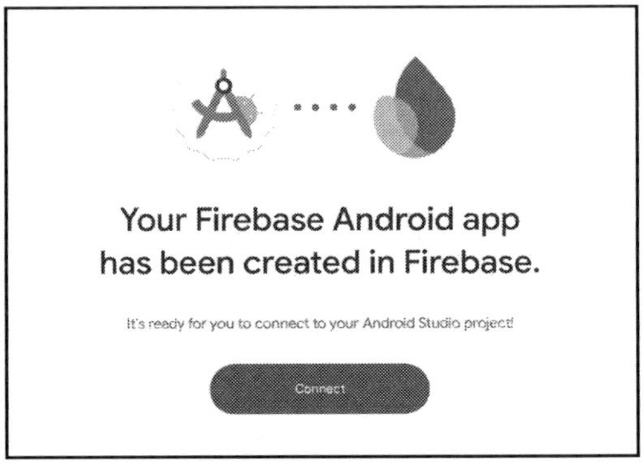

Firebase Connect

6. After you click **Connect**, Android Studio will automatically configure your project. The Firebase console will display a message if the connection is successful.

Firebase Confirmation

Step 4: Review Billing Plan in the Firebase Console

This step is required for image generation. Setting up a billing plan is a prerequisite for using **Imagen API**. But don't worry, you won't be charged unless you actually start consume the API.

> **Note**: Review the Imagen4 Pricing Info (https://ai.google.dev/gemini-api/docs/pricing#imagen-4) for image generation.

1. Open the **Project Overview** page in the Firebase console.
2. By default, the billing plan is set to the **Spark plan**. Tap the plan information to change it to **Blaze plan**.

Default Plan

3. In the **Firebase pricing plans** window, select **Blaze plan**.

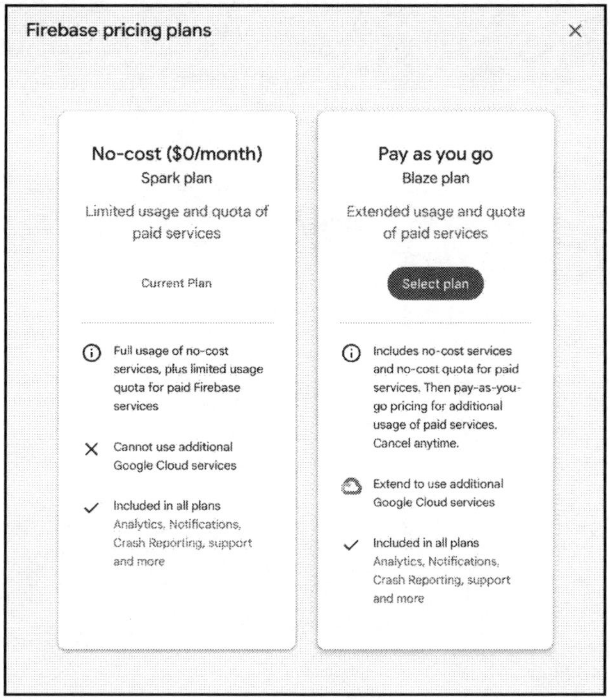

Select Plan

4. Click **Create a Cloud Billing account** and complete the billing setup.

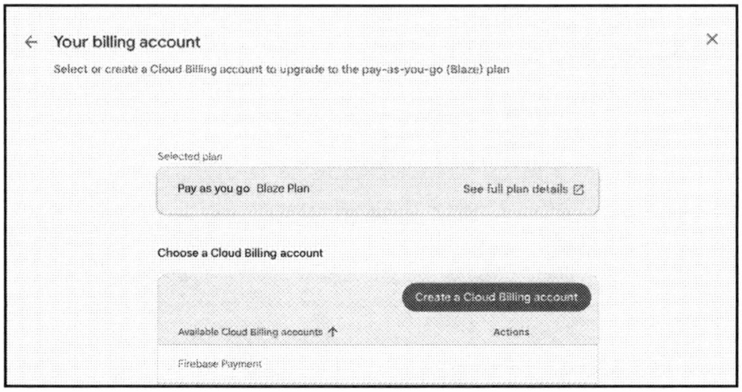

Create a Cloud Billing

5. Before linking the Cloud Billing account, remember to set a billing budget. You can set the budget as low as you prefer — this helps prevent accidental charges while testing your app. Firebase will notify you via email at 50%, 90%, and 100% of your budget usage so you can take preventive action.

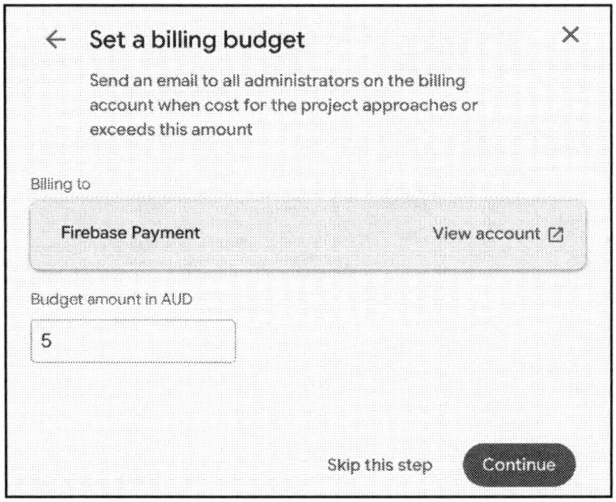

Billing Budget

6. Once your billing account is successfully connected and the project is upgraded to the **Blaze** plan, a confirmation message will appear. Press **Done** to return to the **Project Overview** page.

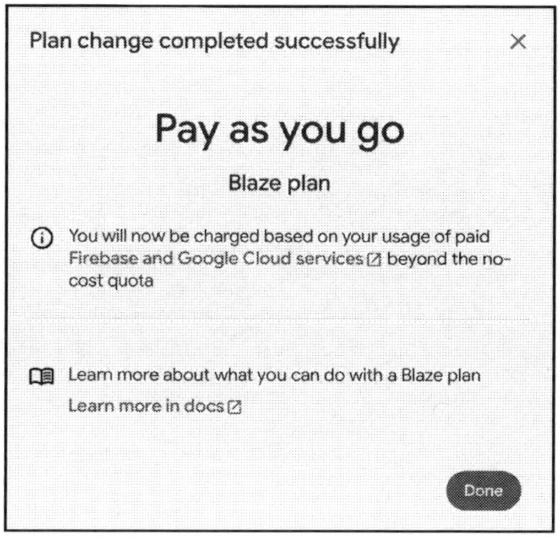

Blaze Plan Successful

Step 5: Enable Gemini Developer API in the Firebase Console

1. From the **Project Overview** page, select `Product categories > AI > AI Logic`. Tap the **Get started** button.

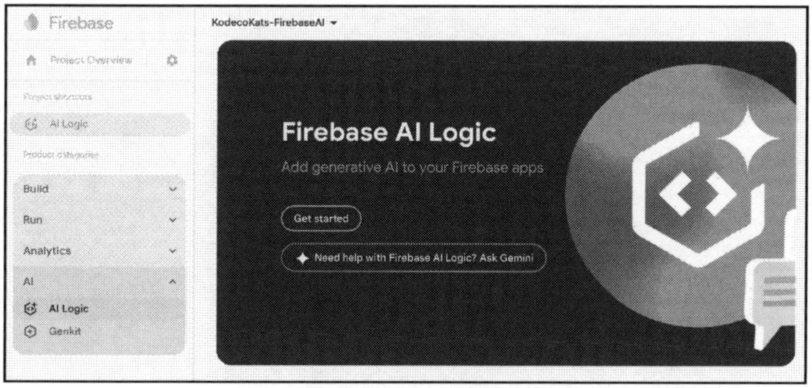

Get Started AI Logic

2. In the next window, select **Get started with this API** under the **Gemini Developer API** panel, and then press **Enable**. This will create API keys and enable the necessary APIs.

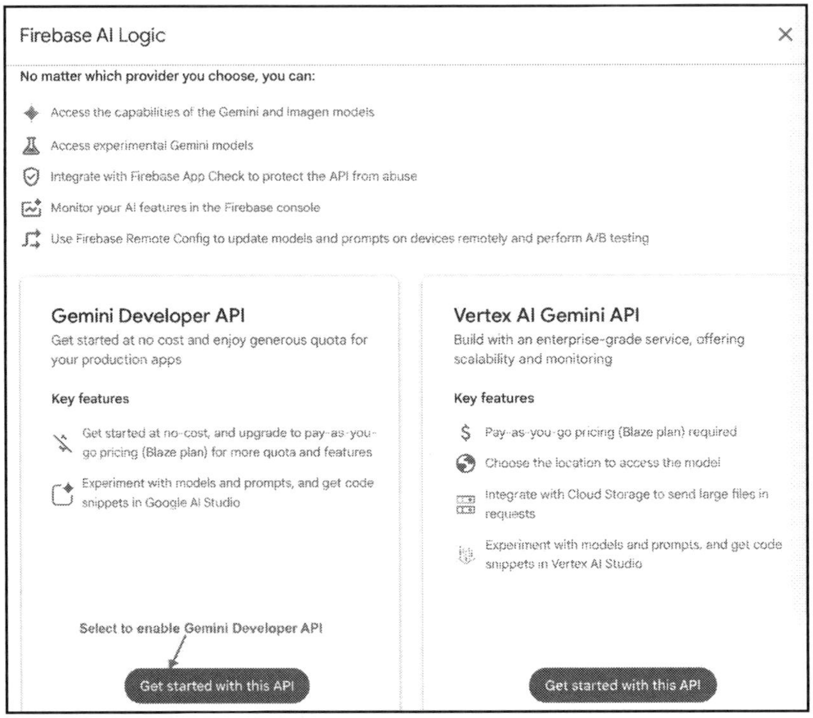

Gemini Developer API

3. Tap **Continue** to finish the setup. You're now ready to consume Firebase AI Logic APIs from the app.

Step 6: Add Firebase AI Logic from Android Studio

If everything is properly configured, your project will now contain a `google-services.json` file inside the **app** directory. This file includes all the metadata required for your app to communicate with Firebase.

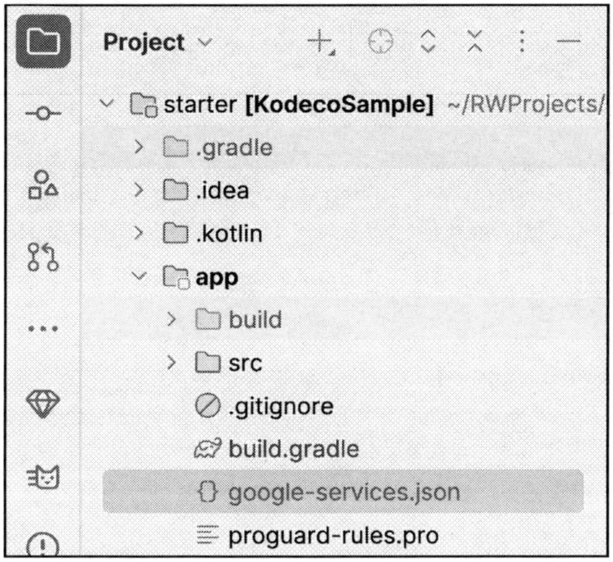

google-services

> **Note**: If you try out the final project (https://github.com/kodecocodes/aam-materials/tree/404b4d117620eee964794e75d85ea6a73ffc7cea/06-harnessing-cloud-power-with-firebase-ai-logic/projects/final) for this chapter, remember to replace the included google-services.json file with your own configuration file.

At this point, you should see a **Connected** status at the top of the Firebase Assistant panel, indicating that Android Studio has successfully linked your project with Firebase.

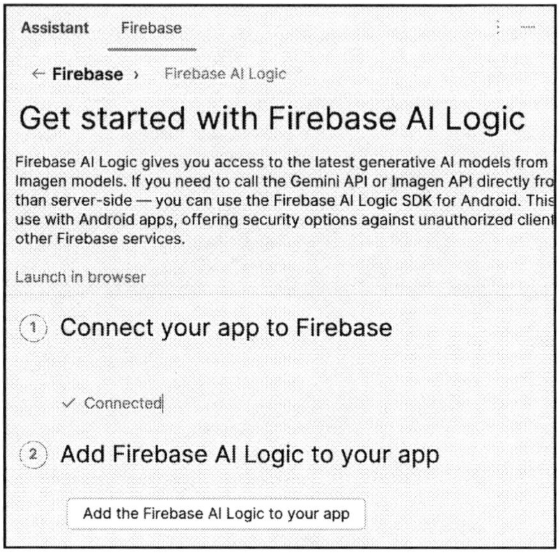

Add Firebase AI Logic

1. Click **Add the Firebase AI Logic to your app** button there.
2. Tap **Accept Changes** in the popup dialog. Android Studio will automatically add the necessary dependencies for Firebase AI Logic.

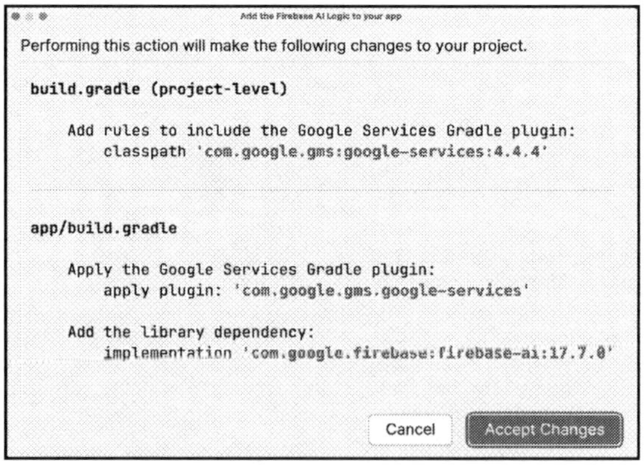

Firebase AI Logic Dependencies

This will sync your project with Gradle and start downloading the required dependencies.

Step 7: Initialize Firebase AI Instance

Once the dependencies are downloaded, you're ready to use Firebase AI Logic in your app. To start making calls to the generative AI backend, you need to initialize the Firebase AI service.

Open the **MainViewModel.kt** and add the following imports at the top of the file:

```
import com.google.firebase.Firebase
import com.google.firebase.ai.ai
import com.google.firebase.ai.type.GenerativeBackend
```

Now, add the following code inside the `MainViewModel` class:

```
val firebaseAI = Firebase.ai(
  backend = GenerativeBackend.googleAI(),
  useLimitedUseAppCheckTokens = true
)
```

The `firebaseAI` instance provides access to the Firebase AI APIs for interacting with **Gemini** and **Imagen** models directly from your app.

- The `backend` parameter defines that you're using Google's **Generative AI** backend, which is the recommended API provider for Android Developers.

- Setting `useLimitedUseAppCheckTokens = true` ensures that Firebase App Check (https://firebase.google.com/docs/app-check/android/play-integrity-provider) is enabled to limit token misuse when sending requests to the backend. Firebase App Check helps prevent API calls from unauthorized clients by verifying the integrity of your app.

Here's how your app interacts with AI models through Firebase AI Logic:

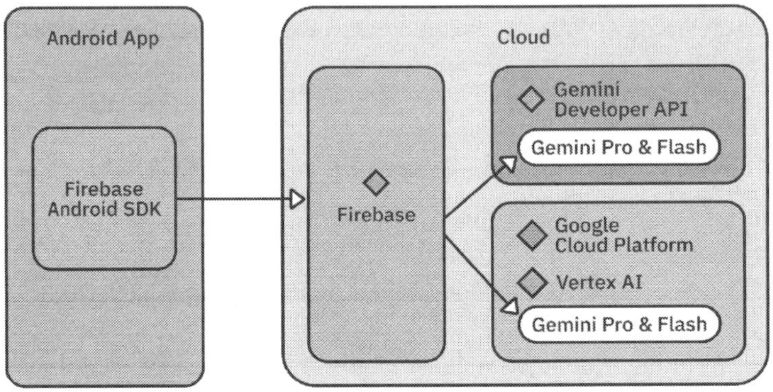

Firebase AI Logic Integration

You're now ready to make your first API call to Firebase AI!

Generating Images with Imagen 4

Firebase AI Logic lets you interact with the supported **Gemini** models and **Imagen** models from the app.

The key difference between Gemini and Imagen models is that Gemini models are multimodal because they can process and generate multiple modalities, such as text, code, PDFs, images, video, and audio. Imagen models, on the other hand, are trained for image generation and modification.

Imagen is best suited for:

- Generating images from natural language prompts.
- Generating images in a wide range of formats and styles.
- Rendering text within generated images.

Choosing Imagen Model

Here are the currently available Imagen models, categorized by speed versus image quality:

Category	Model	Specialty
Imagen 4	imagen-4.0-generate-001	Generates high-quality images from natural language text prompts.
Imagen 4 Fast	imagen-4.0-fast-generate-001	Quickly generates low-fidelity images for prototyping or low-latency use cases.
Imagen 4 Ultra	imagen-4.0-ultra-generate-001	Produces photo-realistic, highest-quality images from text prompts.

When you launch the starter project, you'll see a list of different cat breeds. You're probably curious to know how each breed looks! You'll use **Imagen** to generate an image for each breed you select from the list.

Run the app and select any cat breed of your choice. It'll take you to the `DetailScreen`. Currently, there's a placeholder for cat image and you'll see a loader for generating info.

You'll need to define what sort of image you want to display in the image placeholder - for example, JPEG or PNG, portrait or landscape. The **Imagen API** lets you set these parameters along with the model configuration.

Configuring Output Image

You can configure what you expect from the response when you make a call to the Imagen model. Open **MainViewModel.kt** and define the output configuration as follows:

```
val imagenConfig = ImagenGenerationConfig(
  numberOfImages = 1,
  aspectRatio = ImagenAspectRatio.LANDSCAPE_16x9,
  imageFormat = ImagenImageFormat.png(),
)
```

This block is self-explanatory:

- `numberOfImages` specifies how many images should be generated from a single prompt.

- `aspectRatio` defines the proportions of the generated image (portrait, landscape, or square).
- `imageFormat` determines the file format or compression type (e.g., JPEG or PNG).

If image generation fails, an error will be logged, so watch your **Logcat** console. This can happen for various reasons, for example, no internet connection or if the Firebase project is not configured with a **Blaze plan**.

Defining the Model

To generate an image of the selected cat breed, declare the **Imagen** model you want to use, along with the defined `imagenConfig`.

```
val imagenModel = firebaseAI.imagenModel(
  modelName = "imagen-4.0-fast-generate-001",
  generationConfig = imagenConfig,
)
```

Using `modelName = "imagen-4.0-fast-generate-001"` ensures you're using the fastest and most cost-effective model, since you don't need ultra-high fidelity for this use case.

Generating Image from Prompt

Ready to create an image from a prompt? Add this function below `imagenModel` block in **MainViewModel.kt**:

```
fun generateCatImage(breed: String) {
  val prompt = "A cat of $breed breed."

  viewModelScope.launch(Dispatchers.IO) {
    try {
      _catImage.value = null

      val response = imagenModel.generateImages(prompt)

      val image = response.images.first()
      _catImage.value = image.asBitmap()
    } catch (e: Exception) {
      Log.e("MainViewModel", "Error generating cat image", e)
    }
  }
}
```

This function performs the following steps:

1. Takes the `breed` name as an input parameter and creates a descriptive prompt.
2. Calls `imagenModel.generateImages()` with that prompt to generate an image accordingly.
3. The `response` from Imagen model returns a list of images. As you set `numberOfImages = 1` in the `imagenConfig`, there will be only one item in the list; you're picking the first item from the list.
4. Updates the `_catImage` state by converting the `image` into a Bitmap.

Display the Image

The rest is easy! Go to **DetailScreen.kt** and add following lines inside the `DetailScreen` composable:

```
LaunchedEffect(breedName) {
  viewModel.generateCatImage(breedName)
}
```

The `LaunchedEffect` triggers the `viewModel` to generate an image of the selected cat breed as soon as the Composable is rendered. The `catImage` state automatically updates the `Image` placeholder once the generated image is ready and propagated from `viewModel`.

> Using the **Imagen API** requires your Firebase project to be on the paid "Blaze" (Pay as you go) plan. The Blaze plan includes the same free usage as the Spark plan, plus an additional free tier for paid services like the Imagen API.

Build and run the app, tap on any list item, and navigate to the `DetailScreen` – you'll see a freshly generated image of the selected cat breed appear, courtesy of Imagen!

Image generation using Imagen API

Generating Description with Gemini Model

Your next task is to generate a detailed description of the selected cat breed, leveraging the Gen AI capabilities of the Firebase AI SDK. Getting started is simple, go back to **MainViewModel.kt** and define the AI model:

```
val genAiModel = firebaseAI.generativeModel(modelName =
"gemini-2.5-flash-lite")
```

Here, `genAiModel` gives you access to the `gemini-2.5-flash-lite` model for text generation through `firebaseAI`.

Generating Content from Prompt

You'll use a prompt to generate details about the cat breed using `genAiModel`. To do so, add the following function:

```
fun describeCat(breed: String) {
  val prompt = "Describe the $breed cat."

  viewModelScope.launch(Dispatchers.IO) {
    _isLoading.value = true
    _catDescription.value = "Generating details for $breed
cat..."

    val response = genAiModel.generateContent(prompt)
    val responseText = response.text

    _catDescription.value = responseText ?: "Description not
available."
    _isLoading.value = false
  }
}
```

This function works as follows:

- Creates a `prompt` from provided `breed` name to generate description of that breed.
- Displays a loading state while generating details using `genAiModel`.
- `genAiModel.generateContent(prompt)` does the job – it executes text generation from the text input given to the model as a prompt.
- Updates `_catDescription` with the model's output or a fallback message if generation fails.

Simple as that!

Displaying Content

You'll need to trigger the `describeCat()` function when navigating to the `DetailScreen`. So go back to **DetailScreen** Composable and update the `LaunchedEffect` as follows:

```
LaunchedEffect(breedName) {
  viewModel.generateCatImage(breedName)
  viewModel.describeCat(breedName)
}
```

By adding `viewModel.describeCat()`, you instruct the `viewModel` to generate a textual description for the selected breed.

The `val catDescription by viewModel.catDescription.collectAsState()` statement at the beginning of the `DetailScreen` composable ensures that the description is displayed automatically when ready.

Build and run the app again. Navigate to the `DetailScreen` and you'll see a generated image of the cat along with a detailed description of its breed!

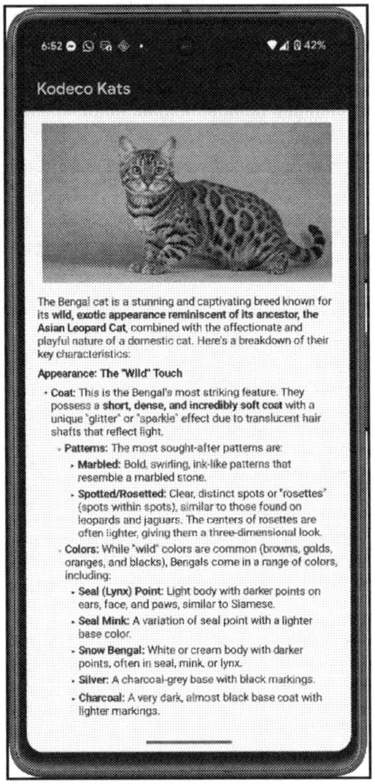

Cat description generation in Detail Screen

Production Readiness Checklist

Deploying an AI feature is not the end of the development process. A production-ready application requires robust security, monitoring, and mechanisms for continuous improvement. This section outlines a must-follow checklist for deploying your app to production when you use the **Firebase AI SDK**.

Secure with Firebase App Check

Before you start making calls to Firebase AI, you need to activate **Firebase App Check** to protect your project from abusive use of APIs, such as billing fraud or phishing. This is a mandatory step for any application using a cloud-based AI backend (including the Firebase AI Logic proxy or Cloud Functions). Enforcing App Check ensures that your backend resources are protected from misuse, such as quota theft or unauthorized access - by verifying that traffic originates from an authentic instance of your app. Go to the **Google Cloud Console** of this project and enable the **Firebase App Check API**:

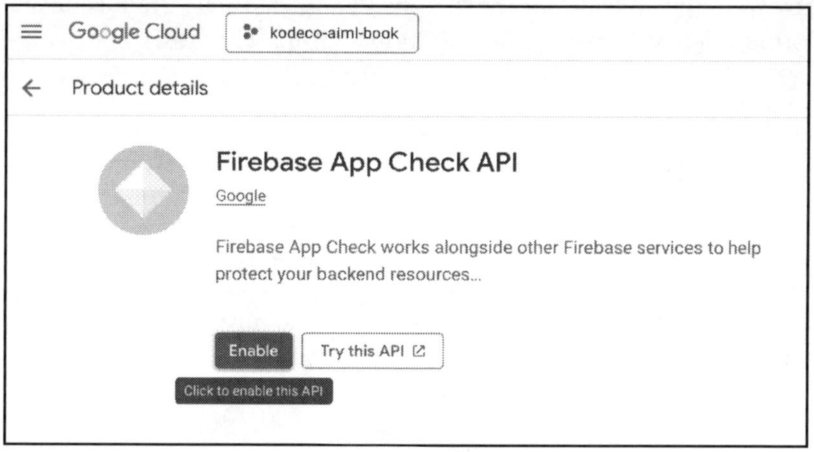

Firebase App Check API

> **Note**: Firebase App Check is strongly recommend when you're using Firebase AI Logic. Visit the official docs (https://firebase.google.com/docs/app-check) for more details on configuring Firebase App Check.

Dynamic Updates with Firebase Remote Config

Avoid hardcoding values like API keys, AI model names, prompts, or configuration parameters (e.g., temperature, top-K) in the app. Instead, store these in **Firebase Remote Config**. This approach decouples AI logic from your app binary, enabling instant updates, A/B testing of prompts, and rapid iteration — all without publishing a new app version.

API Restrictions

In the Google Cloud Console, go to "APIs & Services > Credentials".Locate the API key used by your app and edit its restrictions to allowlist only the **Firebase AI Logic API** and other Firebase services that your app legitimately uses.

Conclusion

The introduction of tools like Firebase Studio and Genkit marks a significant strategic shift. Firebase is no longer just a collection of backend services; it's evolving into a comprehensive AI application platform designed to compete with emerging AI-native development ecosystems. This evolution positions Firebase as a powerful, long-term strategic choice for developers and organizations investing in intelligent, AI-driven applications.

Section III: Advanced Integration, Distribution, and Responsible AI

By this point in your journey, you've explored both the fundamentals of Android AI and the core tools that power intelligent features. Now it's time to move beyond implementation and into the realities of shipping, scaling, and sustaining AI features in production.

In this section, you'll learn:

- How to **package and deliver** on-device ML and GenAI models through the Play ecosystem, enabling dynamic model updates, optimized distribution, and reduced app sizes.
- How to build real-time, multimodal, assistant-like experiences with **Gemini Live**, including streaming audio, session management, and function calling for interactive agents.
- How to **design AI responsibly**, incorporating fairness, transparency, safety, and user control into every part of your app — from data flow to UI.
- How to prepare your AI features for production, covering monitoring, model rollback, budgeting, privacy constraints, and long-term sustainability.
- What the future of Android AI looks like, and how developers can adapt to the rapidly evolving ecosystem.

Across these final chapters, you will not only deepen your technical expertise but also gain the strategic perspective needed to build AI-powered Android apps that scale — ethically, safely, and confidently.

We hope you're ready to jump in and enjoy getting to know the power of AI in Android!

Chapter 7: Optimizing AI Performance & Deployment with Play for On-device AI

Zahidur Rahman Faisal

Hello reader! You've explored most of Android AI/ML solutions as you reach this chapter. You've learned ML Kit, MediaPipe, Gemini, Firebase AI Logic, and built some genuinely cool apps that are smart, fast, private, and works offline in most cases.

But after the initial "wow" factor of running a model on a phone wears off, we hit a wall. It's what I call the "last mile" problem of on-device AI.

The proof-of-concept works great on your high-end test devices. But then the questions start rolling in:

"This new generative model is 500MB. We can't just stick that in the APK, can we?

"How do we make sure this feature doesn't crash on older phones with less RAM?"

"Our data science team has a new, better version of the model. How do we ship it to users without forcing a full app update?"

Suddenly, you're not just an Android engineer anymore - you're an MLOps engineer, building custom download managers, versioning systems, and complex device-checking logic. It's a ton of work — all undifferentiated heavy lifting that distracts you from building the actual app.

For a while, that was just the cost of doing business on the cutting edge. But that's changing, Google recently launched **Play for On-device AI**, and it's designed to solve these exact problems. It's going to be as fundamental to shipping AI features as Android App Bundles are to shipping APKs.

Remember building the On-device LLM app at Chapter 5? You had to download and copy the **TinyLlama** model, which was about 1.25 GB! Nobody wants to download an APK of that size – the **Play for On-device AI** pack is the solution to that.

What Is Play for On-device AI Pack?

Think about the problems that App Bundles solved. Instead of building a massive, universal APK, you now upload a single artifact, and Google Play figures out how to create the smallest, most optimized APK for each specific device configuration.

Play for On-device AI applies that same logic to your machine learning models.

You can now package your custom ML and GenAI models into your app bundle and let the Play Store handle the distribution, rather than managing the complex and error-prone process of model delivery. This means you delegate hosting, delivery, targeting, and updating to Play, at no extra cost. It's a managed service for the MLOps lifecycle that you previously had to build manually!

The best part is that you control when the model is delivered to the user. You can configure model downloads as:

- **Install-time** – delivered with the app during installation.
- **Fast-follow** – downloaded immediately after the app is installed.
- **On-demand** – downloaded only when your app explicitly requests it.

Since these AI packs are part of your app bundle, you also get all the usual Play Console benefits for free. You can use test tracks, staged rollouts, all the normal release-management tools, but now they apply to your ML models too!

Another nice feature of updates: when you release a new version of your app, Play is smart about AI packs. If a particular pack hasn't changed, users won't need to download it again. Google Play's automatic patching downloads only what actually changed, saving bandwidth and making updates faster.

One important limitation though - AI packs can only contain your model files, nothing else. You can't put Java, Kotlin, or native libraries in there. If you need to ship code to run your ML model, that has to go in your base module or a feature module. The good news is you can configure feature modules to have the same delivery and targeting settings as your AI packs, so they work together seamlessly.

Solving the Deployment Puzzle: AI Packs and Delivery Modes

The core concept is the **AI pack**. It's a special container within your app bundle that holds your models. The magic is in how these packs get delivered to the user. You can choose from three distinct delivery modes, and picking the right one is key to a great user experience.

1. **Install-time Delivery**

This is the most straightforward option. The AI pack is delivered and installed alongside the main app, just like a regular APK.

- **When to use:** For critical, must-have AI features. If a core part of your app depends on a model, make sure it's there the first time the user opens the app.
- **Trade-off:** Increases the initial download size from the Play Store. For large models, this can deter users.

2. **Fast-Follow Delivery**

This is the sweet spot for many features. The app itself is installed quickly, and the AI pack begins downloading automatically in the background right after.

- **When to use:** Ideal for important but non-critical AI features. The user can explore your app immediately, and the AI feature "lights up" a few moments later.
- **Trade-off:** You must handle the state where the model isn't ready yet. Show a loading UI or disable the feature until download completes. Your code should check download status on each app launch.

3. **On-Demand Delivery**

This gives you the most control. The AI pack is only downloaded when your app explicitly requests it at runtime.

- **When to use:** Best for large models powering niche features — for instance, a "pro" photo filter or a specialized generative-AI tool used by few users.
- **Trade-off:** Requires more UI work — you must trigger the download, display progress, handle errors, and enable the feature when ready.

By offering these three modes, Play lets you balance feature availability, app size, and user experience intelligently.

An Overview of the Play for On-device AI Workflow

Using **Play for On-device AI** to package and deploy AI Packs is pretty straightforward. Let me break down the process for you:

1. You'll need to package your trained models into AI packs and include them in your Android App Bundle. During this step, specify the delivery mode - **install-time**, **fast-follow**, or **on-demand** delivery for each pack.
2. If you choose **fast-follow** or **on-demand** delivery, you'll need to integrate the **Play AI Delivery Library** into your app. This library handles the actual downloading of your AI packs when needed. If you went with install-time delivery, you can skip this step since the models will already be there when your app launches.
3. If you want more precise targeting, configure which devices receive which models. This step is *optional*, yet it's quite handy! Imagine delivering a lightweight model to budget phones and a more sophisticated model to flagship devices with 6GB+ RAM. You can even target specific device models or exclude devices that can't handle the model.
4. Finally, test and release your app bundle through the Play Console, just like any other update. You can use internal testing, staged rollouts, and every release-management tool you already know.

Later in this chapter, you'll learn how to configure your app for **on-demand** delivery of an ML model, extending the "On-device LLM" project from Chapter 5.

Download the starter project (https://github.com/kodecocodes/aam-materials/tree/editions/1.0/07-play-for-on-device-ai/projects/starter) and open it in Android Studio to begin.

Note that, the starter project contains some convenience methods in the `MainViewModel`, which won't compile until you add the **Play AI Delivery Library** to the project. This is intentional so the chapter can focus on enabling Play-based-on-device capabilities rather than writing boilerplate code. Once you add the Play AI Delivery Library as a dependency (in Step 2), the project will compile and run successfully.

> **Note**: The **TinyLlama** LLM model used in this sample project comes from the **Hugging Face** community and can be downloaded here (https://huggingface.co/litert-community/TinyLlama-1.1B-Chat-v1.0/resolve/main/TinyLlama-1.1B-Chat-v1.0_multi-prefill-seq_q8_ekv1280.task).

Step 1: Configuring Your On-Demand AI Pack

- **Create the AI Pack module:** In the root directory of the Android project, create a new directory for the AI Pack. The directory name will be used as the AI Pack's identifier. It must start with a letter and can only contain letters, numbers, and underscores (e.g., `myModelPack` or `my_model_pack_1`).

For this project, name it `onDemandAiPack`. The project structure should look like this:

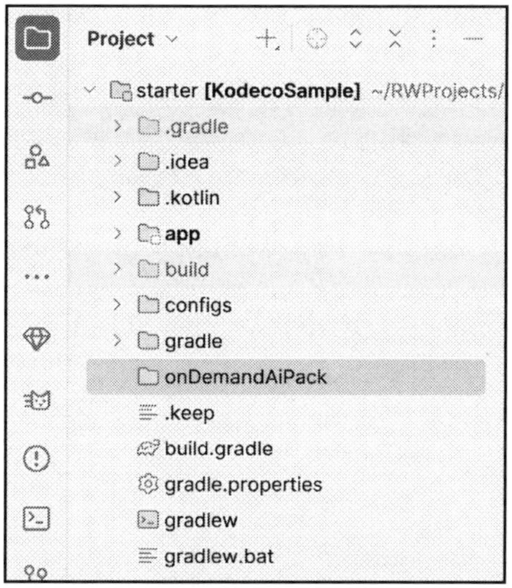

Project Structure

- **Create the module's `build.gradle`:** Inside the `onDemandAiPack` directory, create a **build.gradle** file. In this file, apply the `com.android.ai-pack` plugin, then define the AI Pack's name and specify a delivery mode (`install-time`, `fast-follow`, or `on-demand`).

```
// In onDemandAiPack/build.gradle
plugins {
  id 'com.android.ai-pack'
}
```

```
aiPack {
  packName = "onDemandAiPack"
  dynamicDelivery {
    deliveryType = "on-demand"
  }
}
```

- **Register the pack in the app module:** In the main application module's **build.gradle** file (typically `app/build.gradle`), declare the new AI Pack inside the `android` block.

```
// In app/build.gradle
android {
  //...
  assetPacks = [":onDemandAiPack"]
}
```

- **Include the module in `settings.gradle` file:** Ensure the Gradle build system recognizes the new module by adding an `include` statement in the project's root **settings.gradle** file.

```
// In settings.gradle
include ':app'
include ':onDemandAiPack'
```

- **Structure the AI Pack and place the model:** Within the AI Pack module directory, **onDemandAiPack**, create the standard source-set structure: `src/main/assets/`. Place your AI model file (for example, *TinyLlama-1.1B-Chat-v1.0_multi-prefill-seq_q8_ekv1280.task*) inside the `assets` directory.

This path is preserved during the build and deployment process, and the final accessible path on the device will look like: `[pack_path]/assets/[model_file_name]`

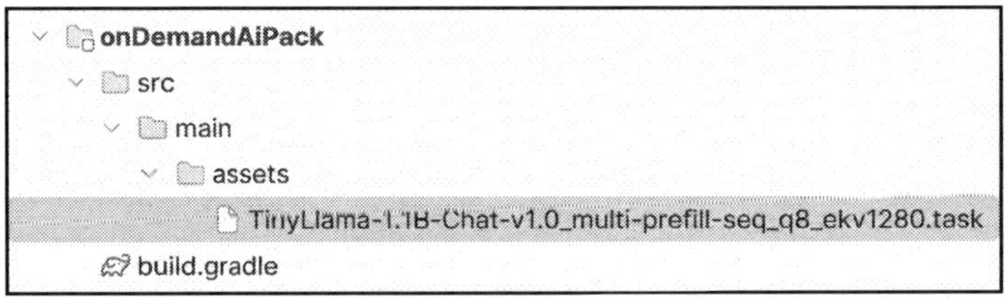

AI Pack Structure

Step 2: Integrating and Using Play AI Delivery

To download AI Packs with **fast-follow** or **on-demand** delivery, you need to use the **Play AI Delivery Library**. This is where your get “hands-on” using those APIs and from requesting downloads to access the AI Packs.

The process includes a few steps:

- Including the library dependency.
- Checking any existing or downloaded AI Pack status.
- Adding capability to download the AI Packs, or cancel any ongoing downloads.
- Monitoring status updates of the downloads and displaying them on the UI.
- Finally, accessing or using the AI Packs once it’s downloaded.

Adding the Dependency

Get started by adding this dependency In the app-level **build.gradle** file:

```
// In app/build.gradle
dependencies {
  ...
  // TUTORIAL DEPENDENCIES
  var aiDeliveryVersion = "0.1.1-alpha01"
  implementation "com.google.android.play:ai-delivery:
$aiDeliveryVersion"
}
```

Sync the project with Gradle files from the Android Studio. Open **MainViewModel.kt** and import all the necessary libraries.

Build and run, the project should compile successfully. At this point, you'll see a loading state: "Checking for AI Model...".

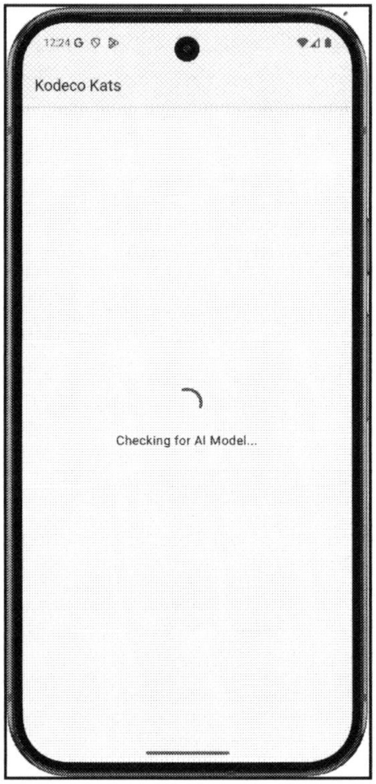

Checking AI Model

Now you need to complete the status check and update the UI with results.

Checking the Status of AI Packs

Before using assets or models from an AI pack, you should check whether the pack has already been downloaded.

Open **MainViewModel.kt**, the `aiPackManager` serves as the primary interface for managing the download and state of AI Packs delivered via fast-follow or on-demand modes.

Now, update the `checkPackStatus()` function as follows:

```
fun checkPackStatus(packName: String) {
  val packLocation = aiPackManager.getPackLocation(packName)
  if (packLocation != null) {
    _aiPackStatus.value =
AiPackStatus.Installed(packLocation.toString())
    return
  }

  aiPackManager.getPackStates(listOf(packName))
    .addOnSuccessListener { states ->
      val state = states.packStates()[packName]
      _aiPackStatus.value = mapAiPackStatus(state = state)
    }.addOnFailureListener { e ->
      _aiPackStatus.value = mapAiPackStatus(state = null)
    }
}
```

What it does is straightforward:

- Takes a `packName` to check if it's already installed. If so, it'll return the asset's location.
- Otherwise, `aiPackManager` will invoke `getPackStates()` function. You can pass a list of packages to that function to check states sequentially and check state of the given `packName` here. On a successful completion, you need to map the state to local states and update the UI accordingly.
- In case of failure, the status is unknown, and you need to reflect that on the UI. Setting `_aiPackStatus.value = mapAiPackStatus(state = null)` does that.

But why `mapAiPackStatus()` function is needed? To understand, you need to have a deeper look at that; You'll explore that shortly.

Fetching AI Packs

The `aiPackManager` handles the heavy-lifting. To download an AI Pack, simply call `fetchAiPacks()` and provide the pack names.

Add `fetchAiPacks()` function in the **MainViewModel** to do so.

```
fun fetchAiPacks() {
  aiPackManager.fetch(AI_PACKS)
}
```

Here, `AI_PACKS` is a list containing the names of all AI Packs you want to download.

Cancelling Requests

In case you want to cancel any fetch request, cancelling is as simple as fetching AI Packs using Play AI Delivery Library. Adding this function to the **MainViewModel** will allow you to do that:

```
fun cancelRequests() {
  aiPackManager.cancel(AI_PACKS)
}
```

The `aiPackManager.cancel()` method can take a list of AI Packs and cancels any ongoing downloads for those.

Displaying Status Updates

Play AI Delivery Library returns `AiPackState` while fetching AI Packs or checking AI Pack status. The `AiPackState` is a public `interface` that looks like this:

```
@Retention(RetentionPolicy.CLASS)
public @interface AiPackStatus {
  int UNKNOWN = 0;
  int NOT_INSTALLED = 8;
  int PENDING = 1;
  int WAITING_FOR_WIFI = 7;
  int REQUIRES_USER_CONFIRMATION = 9;
  int DOWNLOADING = 2;
  int TRANSFERRING = 3;
  int COMPLETED = 4;
  int FAILED = 5;
  int CANCELED = 6;
}
```

To translate them into a simpler from for the user, you have an `AiPackStatus` data class within the **aipack** package in the sample project. Let's look at how the mapping works in the `mapAiPackStatus()` function inside **MainViewModel**:

```
private fun mapAiPackStatus(state: AiPackState?): AiPackStatus {
  return when (state?.status()) {
    AssetPackStatus.NOT_INSTALLED -> {
      AiPackStatus.NotInstalled
    }

    AssetPackStatus.WAITING_FOR_WIFI,
    AssetPackStatus.REQUIRES_USER_CONFIRMATION -> {
      AiPackStatus.RequestConfirmation
    }

    AssetPackStatus.PENDING,
    AssetPackStatus.DOWNLOADING,
    AssetPackStatus.TRANSFERRING -> {

AiPackStatus.Downloading(state.transferProgressPercentage())
    }

    AssetPackStatus.COMPLETED -> {
      val assetLocation = getAssetFromAiPack(
        packName = AI_PACK_NAME,
        assetName = Model.TINYLLAMA_1_1B_CHAT_V1_0.modelName
      )
      if (assetLocation == null) {
        AiPackStatus.NotInstalled
      } else {
        AiPackStatus.Installed(location = assetLocation)
      }
    }

    AssetPackStatus.CANCELED,
    AssetPackStatus.FAILED -> {
      AiPackStatus.Failed(errorCode = state.errorCode())
    }

    else -> {
      AiPackStatus.Unknown
    }
  }
}
```

A few things to note for the above function:

1. If the pack isn't installed yet, it displays the `AiPackStatus.NotInstalled` state.

2. The next block handles cases where user interaction is required:

- WAITING_FOR_WIFI means the download is paused because the user is not on Wi-Fi.
- REQUIRES_USER_CONFIRMATION means Google Play requires the user to explicitly approve the download of the requested asset.

In either case, you map it to the RequestConfirmation state, which tells the UI to show a dialog.

3. AssetPackStatus.PENDING, AssetPackStatus.DOWNLOADING and AssetPackStatus.TRANSFERRING – these three states all indicate that the download is in progress. You map them to a single Downloading state and pass the progress percentage to the UI layer.
4. AssetPackStatus.COMPLETED is the "success" case. It means the pack is downloaded on the device. The Installed state uses getAssetFromAiPack() helper function to obtain the actual file path of the model. If for some reason the path isn't available (e.g., the asset wasn't downloaded properly), it's treated as NotInstalled state.
5. If the download was CANCELED by the user or FAILED for some reason, the AiPackStatus.Failed state is displayed along with the errorCode, so you can log it or show a more specific message.
6. AiPackStatus.Unknown is the fallback state. It helps prevent crashes is case of any unexpected statuses.

Listening to Status Updates

Whenever you fetch an AI Pack or check status, listening to these status updates is made easy by the Play AI Delivery Library – you need to use the AiPackStateUpdateListener interface from the library.

For your convenience, the **MainViewModel** implements AiPackStateUpdateListener. All you need to do is as follows:

Attach the listener to aiPackManager within the init function of **MainViewModel**.

```
init {
  aiPackManager.registerListener(this)
}
```

Override the `onStateUpdate()` function to map AI Pack states to your own `AiPackStatus`.

```
override fun onStateUpdate(state: AiPackState) {
  Log.d(TAG, "onStateUpdate: $state")
  _aiPackStatus.value = mapAiPackStatus(state)
}
```

Accessing AI Packs

Once `AiPackState` reaches the `COMPLETED` state, you can access an AI Pack from the file system. Calling `aiPackManager.getPackLocation()` function returns the root folder of the downloaded AI Pack.

AI Packs (such as models) are stored in the assets directory within the root directory. You can retrieve the absolute path of a specific asset file as follows:

```
private fun getAssetFromAiPack(packName: String, assetName:
String): String? {
  val aiPackLocation = aiPackManager.getPackLocation(packName)
  val assetsFolderPath = aiPackLocation?.assetsPath()
  val assetFile = File(assetsFolderPath, assetName)
  Log.d(TAG, "Asset path: ${assetFile.absolutePath}")

  return if (assetFile.exists()) assetFile.absolutePath else
null
}
```

Here, `assetFile.absolutePath` is the full usable path for the MediaPipe ML inference engine to load the LLM Model.

[Optional] Step 3: Slaying the Fragmentation Dragon by Device Targeting

Remember the "last mile" theory at the beginning of this chapter? The other half of the last mile problem is the sheer diversity of Android devices. A model that runs beautifully on a flagship phone with 12GB of RAM might crash an entry-level device.

Historically, the options were limited – you could either ship a "one-size-fits-all" model that's underpowered for high-end devices, or you could write a ton of custom logic to try and guess the device's capabilities.

Play for On-device AI solves this with model variants and targeting. You can create different versions of your model optimized for different hardware, and Play will automatically deliver the right one.

You can create targeting rules based on device properties like:

- Device RAM.
- System on Chip (SoC).
- Specific device models.
- System features.

You can enable Device Targeting with the following steps:

1. Enable the device targeting feature in the Android Gradle Plugin by adding this to **gradle.properties** file:

```
android.experimental.enableDeviceTargetingConfigApi=true
```

2. Create a device targeting configuration XML file. Define your custom **device groups** in that file. For example, you can create a device group called `gemini_on_device` that targets all devices powered by the **Tensor G5** chip:

```
<config:device-targeting-config xmlns:config="http://
schemas.android.com/apk/config">

  <config:device-group name="gemini_on_device">
      <config:device-selector>
          <config:system-on-chip manufacturer="Google"
model="TensorG5"/>
      </config:device-selector>
  </config:device-group>

</config:device-targeting-config>
```

3. Include your device targeting configuration file in your app bundle by adding this to your main module's `build.gradle` file:

```
android {
  ...
  bundle {
    deviceTargetingConfig = file('device_targeting_config.xml')
    deviceGroup {
      enableSplit = true   // split bundle by #group
      defaultGroup = "other"  // group used for standalone APKs
    }
  }
  ...
}
```

You can package all these variants into the app bundle, define your own targeting rules, and Play handles the rest. Every user gets the best possible experience their hardware can support, without us writing a single line of device-specific if/else logic!

> Device Targeting is *optional* and not used in the sample project to allow it run on most configurations, even on the Emulators. For more details on Device Targeting, you may check official Device Targeting Configuration (https://developer.android.com/google/play/on-device-ai#device-targeting-configuration) documentation.

Step 4: Testing On-Device AI Packs Locally

Testing on-demand features (such as AI Packs with models) traditionally requires uploading builds to the Google Play Console - a process that introduces significant latency into the development cycle, and that can be the trickiest part of the workflow.

Alternatively, you can generate APKs with AI Packs using bundletool and sideload them onto your device.

In this section, you'll learn the step-by-step command-line operations required to build the application and deploy it for local AI Pack testing using bundletool.

> **Note**: Before you begin, ensure you have Java Development Kit (JDK) and JetBrains Runtime (JBR) (https://developer.android.com/build/jdks) installed and configured on your machine.

What is Bundletool?

Bundletool is the underlying tool that Android Studio, the Android Gradle plugin, and Google Play use to build an Android App Bundle. It's also available as a command-line tool (https://developer.android.com/studio/command-line/bundletool).

Bundletool converts your app bundle into the APKs that would be installed on devices, replicating how Google Play generates and serves APKs — including those with AI Packs. This makes it possible to locally test and simulate the deployment process of your app with AI Packs before releasing it to users.

To begin with, download the latest release (https://github.com/google/bundletool/releases) of bundletool (`.jar`). Ensure you have a test device or Emulator connected.

If you're on Mac or Linux, you can install using this homebrew (https://brew.sh) command:

```
brew install bundletool
```

The Core Workflow: Using Bundletool for On-Device Deployment

In this section, you'll generate signed App Bundles or APKs, which require your keystore information – just like preparing an app for deployment to the Google Play Store.

You can initiate this process through the Android Studio menu:`Build > Generate Signed Bundle or APK...`

You may create a new keystore by selecting **Create new...** and providing required information on the next screen as follows:

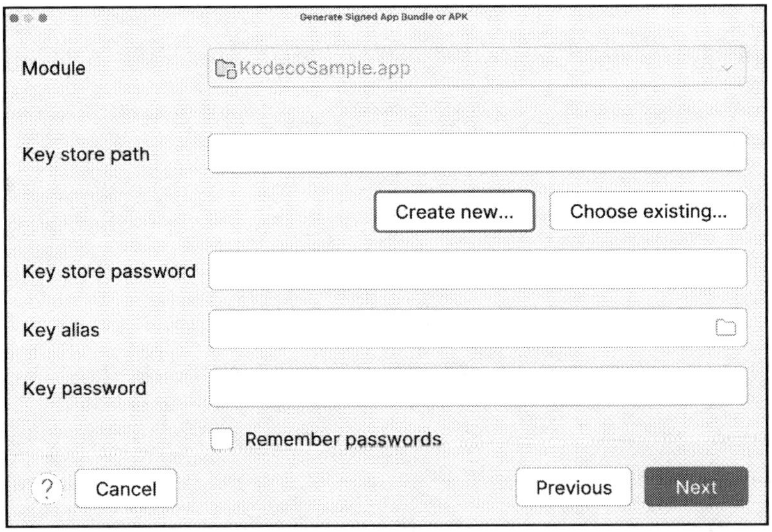

Existing Keystore

Create New Keystore

Or, if you already have an existing keystore file, simply enter your credentials and tap **Next**.

Once your keystore is configured, open the terminal and run the following commands step by step.

> **Note**: The default `debug.keystore` will be used when you run these commands.

1. The workflow begins with creating the standard, **release-signed** Android App Bundle. You can use this terminal command (or use the **Build** menu) in the Android Studio:

```
./gradlew :app:bundleRelease
```

The resulting **.aab** file, typically located at `app/build/outputs/bundle/release/app-release.aab` directory, serves as the primary input for the subsequent bundletool commands.

2. Generate the local-testing APK set (`.apks`) with the command below:

```
bundletool build-apks \
--bundle=app/build/outputs/bundle/release/app-release.aab \
--output=app/build/outputs/bundle/release/app-release.apks \
--local-testing \
--connected-device
```

This command processes the `.aab` file and generates a set of split APKs. It also injects the necessary local testing metadata (using the `--local-testing` flag), and packages everything into a single `app-release.apks` output file.

Upon successful execution, the **outputs** folder will look like this:

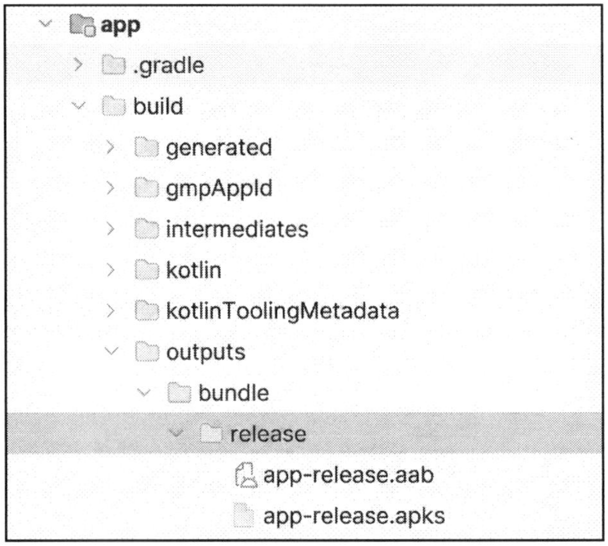

Build Folder

3. Deploy the app and sideload the AI Pack assets with the below command:

```
bundletool install-apks --apks=app/build/outputs/bundle/release/
app-release.apks
```

This command determines the required base and configuration APKs for the connected device and installs them.

Uninstall any previous version of the app from the device before executing this command, as the local-testing workflow does not support in-place updates. When complete, the application icon appears on the device, ready for testing.

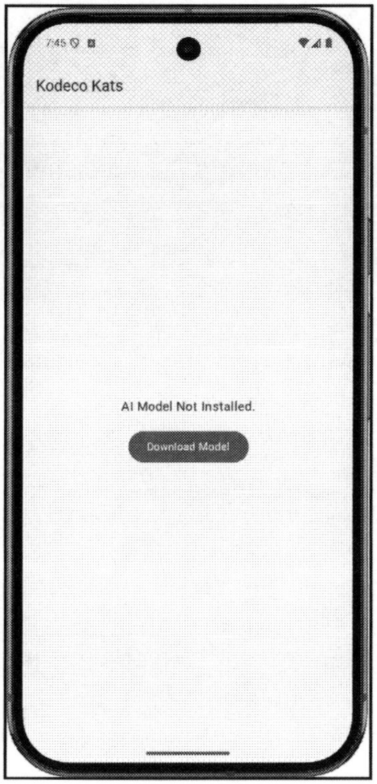

Ready for Testing

Build and run the app and tap the **Download Model** button, verify that the download process and status updates work as expected.

Conclusion

Shipping on-device AI has always been a battle on two fronts: building the feature and building the infrastructure to deploy and manage it. Play for On-device AI effectively eliminates that second front.

The end-to-end process can be summarized as follows:

1. **Configure:** Set up the AI Pack as a distinct Gradle module with an **on-demand** delivery type and place the model file in its assets directory.
2. **Build:** Generate a signed Android App Bundle using `./gradlew :app:bundleRelease`.
3. **Generate APKs:** Use `bundletool build-apks` with the `--local-testing` flag to create **.apks** file.
4. **Deploy:** Uninstall any previous app version, then use `bundletool install-apks` to install the app and sideload AI Pack assets.
5. **Test:** Launch the app and verify that the code requests the pack via `AiPackManager` and accesses the model file using the path provided by `AiPackLocation`.

By handling the hosting, delivery, updating, and targeting of your models, the **Play for On-device AI Library** lets you offload the MLOps complexity to the Play Store. This means smaller app size, more optimized app updates, a better user experience tailored to each device, and most importantly, more time for you to focus on what you do best – building amazing, intelligent apps!

Chapter 8: Building Interactive App with Gemini Live

Zahidur Rahman Faisal

Have you checked out the **Gemini Live API**? It's a total game-changer for building real-time, interactive experiences in Android. Forget managing a whole backend just to stream audio or video to an LLM — Gemini Live makes it effortless.

Imagine building an app where the user can talk to a chatbot and get instant responses, just like a real conversation. That's what the Live API enables.

What Makes an App 'Interactive'?

When we talk about an interactive app in this context, especially with the **Gemini Live API**, we're talking about an application that doesn't just listen and reply — it actually acts on the user's instructions. It goes beyond a simple question-and-answer chatbot.

Think of it this way:

- **Standard Chatbot App:** You say, "What's the weather like?" The model figures out the answer and replies. That's a back-and-forth conversation.

The key is that the user's voice prompt is translated directly into app-logic execution. The app is no longer just a passive interface; it's an agent that can manipulate its own data and features based on a natural language command. It creates a seamless, hands-free experience where the AI is integrated directly into the core functionality of the app — that's what makes it truly 'interactive' in the most powerful sense.

The Gemini Live API

When I first worked with the Gemini Live API, I realized it's a major leap for mobile generative AI. Instead of the old request–response model, it now supports real-time, two-way streaming. That means the client and model can send and receive data simultaneously — creating a **live conversation** rather than a sequence of turns.

It provides an optimized low-latency stream for both the audio you send to the model (your speech as requests) and the audio/text the model sends back (its response). Since you've used **Firebase AI Logic** in earlier chapters, you can invoke this directly from your Android app — with no need for a custom server. It's essentially a bidirectional real-time audio channel connecting straight to a Gemini model.

With this, you can build conversational apps that feel more like talking to another person than interacting with a bot. Users can even interrupt Gemini mid-sentence, and the model smoothly adjusts. The API handles continuous multimodal input (currently audio + text) and outputs immediate speech, which is perfect for voice-first or hands-free interfaces.

Gemini Live has built-in support for:

- Robust session management for long conversations.
- Function calling or tool use
- Voice Activity Detection (VAD) to detect pauses in speech
- Multimodal streaming in both directions

> **Note**: Firebase AI Logic for Gemini Live API is currently in developer preview, meaning that non-backward compatible changes may occur in future releases.

Hands On Gemini Live

Let's extend the Firebase AI Logic app from the previous chapter with Gemini Live bidirectional streaming.

Download the starter project (https://github.com/kodecocodes/aam-materials/tree/editions/1.0/08-building-interactive-app-with-gemini-live/projects/starter), and open it in Android Studio.

Project Setup and Dependencies

First things first, ensure you're targeting Android API level 23+ and the app is connected to Firebase.

Open the app level **build.gradle** file, add the Firebase AI Logic at the end of `dependencies` block.

```
// Firebase AI Logic: Gemini Live Dependency
var firebaseAiLogicVersion = "17.6.0"
implementation "com.google.firebase:firebase-ai:
$firebaseAiLogicVersion"
```

Since you'll be interacting with Gemini Live through **audio**, add this permission in **AndroidManifest.xml**:

```
<uses-permission android:name="android.permission.RECORD_AUDIO"/
>
```

Now, sync your project with the Gradle files and verify that it compiles without any errors.

Build and run the app, and select an item from the list. On the item-detail screen, you'll see button at the bottom showing:

`"WARNING: Gemini Live Not Initialized"`

Before Initialization

Model Initialization and Configuration

The first step in using Gemini Live is initializing the backend service and creating a `LiveGenerativeModel` instance. The Live API configuration is handled through the `liveGenerationConfig` object, which determines the model's behavior and the nature of the streaming output.

To handle initialization cleanly, the best practice is to create a utility/manager class. The starter project already has that for your convenience.

Go to the **com.kodeco.android.aam.live** package and open the `LiveModelManager` file. This handles interacting with the Gemini Live model and updates UI states. These variables below are declared within the class for the same reason:

```
// The core Gemini Live model instance.
private lateinit var liveModel: LiveGenerativeModel

// Mutable state flow holding the current state of the live
session.
private val _liveSessionState =
MutableStateFlow<LiveSessionState>(LiveSessionState.Unknown())
val liveSessionState = _liveSessionState.asStateFlow()
```

Now, add the `initializeGeminiLive()` function as follows:

```
fun initializeGeminiLive(activity: Activity) {
  requestAudioPermissionIfNeeded(activity)

  coroutineScope.launch {
    try {
      val liveGenerationConfig = liveGenerationConfig {
        speechConfig = SpeechConfig(voice = Voice("FENRIR"))
        responseModality = ResponseModality.AUDIO
      }

      liveModel = Firebase.ai(backend = googleAI()).liveModel(
        modelName = "gemini-live-2.5-flash-preview",
        generationConfig = liveGenerationConfig,
      )

      _liveSessionState.value = LiveSessionState.Ready()
    } catch (e: Exception) {
      _liveSessionState.value = LiveSessionState.Error(message =
e.localizedMessage)
    }
  }
}
```

This function requests audio permissions, sets up the configuration, and prepares the model for use.

When configuring the model, several parameters are mandatory for building with Gemini Live:

- **Model Selection:** The `modelName` parameter must explicitly reference a **Live API** compatible model, such as `gemini-live-2.5-flash-preview`.

> Here's the list (https://firebase.google.com/docs/ai-logic/live-api?api=dev#models-that-support-capability) of models that supports the Live API.

- **Output Modality:** The `responseModality` parameter must be set to `ResponseModality.AUDIO` to generate and stream *Text-to-Speech (TTS)* output back to the client.

- **Voice Configuration:** The `speechConfig` parameter allows for specifying a preferred voice, for example, `Voice("FENRIR")`.

> Pro Tip: You can also provide an additional parameter, `systemInstruction` for defining the model's persona, its role (e.g., "You are a helpful assistant, an expert about domestic pets."), and establishing clear operational constraints.

The `initializeGeminiLive()` function is relaying the 'Ready' state after initializing the `liveModel`.

```
_liveSessionState.value = LiveSessionState.Ready()
```

It also throw an 'Error' with a message in case of failure:

```
_liveSessionState.value = LiveSessionState.Error(message =
e.localizedMessage)
```

The `LiveSessionState` is a sealed interface, containing data classes that reflects current state of the Gemini Live session. The `LiveSessionState` is located in the same package and defined as below:

```
sealed interface LiveSessionState {
  data class Unknown(val message: String = "UNKNOWN: Gemini Live
Not Initialized") : LiveSessionState

  data class Ready(val message: String = "READY: Ask Gemini
Live") : LiveSessionState

  data class Running(val message: String = "RUNNING: Gemini Live
Speaking...") : LiveSessionState

  data class Error(val message: String = "ERROR: Failed to
initiate lGemini Live") : LiveSessionState
}
```

This is used to notify updates of the session in the screen.

The UI interactions are managed by **MainViewModel** class, which has a `LiveModelManager` instance as shown below:

```
private val liveModelManager = LiveModelManager(
  context = application,
  coroutineScope = viewModelScope,
)
```

It also contains `liveSessionState` to observe any state changes in `liveModelManager`.

```
val liveSessionState = liveModelManager.liveSessionState
```

Add the below function that'll initialize the Gemini Live model:

```
fun initializeGeminiLive(activity: Activity) {
  liveModelManager.initializeGeminiLive(activity)
}
```

You need to call this function from the **MainActivity** when your app launches.

Now go to the **MainActivity.kt**, add the below code inside the `onCreate(savedInstanceState: Bundle?)` block:

```
viewModel.initializeGeminiLive(activity = this@MainActivity)
```

Build and run again, you'll see a permission dialog for recording audio:

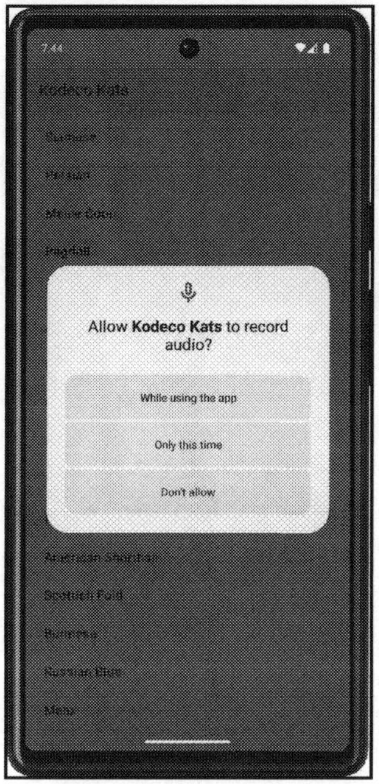

Audio Permission

Approve the permission, and then select an item from the list to go to the next screen.

You'll see the Gemini Live button is now displaying the **READY** state!

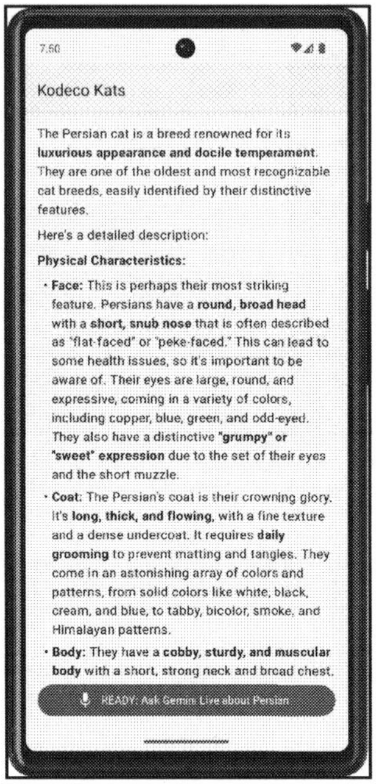

Model Initialized

Real-Time Connection: Starting The Live Session

At this point, the app can connect to Gemini and start the live session. You need to use `LiveModelManager` for that.

Open the `LiveModelManager` class, declare a `LiveSession` instance.

```
private var session: LiveSession? = null
```

The `LiveSession` object is the core abstraction for continuous interaction. It represents the persistent connection established with the model and manages all input/output streaming.

The Live API is bidirectional, and you'll use it for:

1. Sending Voice commands.
2. Generating speech from text.
3. Transcribing audio to text.

Eventually, you will also be able to send images and a live video stream to the model. In the sample app, you'll implement **text -> speech** in this flow:

`List of Cat Breeds > Select a breed > Gemini Live Describes it`

To do so, add the `startSessionFromText()` function in the `LiveModelManager`:

```
@RequiresPermission(Manifest.permission.RECORD_AUDIO)
fun startSessionFromText(catBreed: String) {
  val text = "Tell me about $catBreed cats in maximum 80 words."

  coroutineScope.launch(Dispatchers.IO) {
    try {
      // Start the conversation
      session = liveModel.connect()
      session?.send(text)
      session?.startAudioConversation()

      // Update State
      _liveSessionState.value = LiveSessionState.Running()
      } catch (e: Exception) {
        _liveSessionState.value = LiveSessionState.Error(message
= e.localizedMessage)
    }
  }
}
```

This function accepts a single string argument, `catBreed` that you have selected, and then executes as described below:

1. `val text = "Tell me about $catBreed cats."`: It constructs a text prompt using the input `catBreed`. For example, if "Siamese" is passed to the function, then `text` will be *"Tell me about Siamese cats in maximum 80 words."*

2. `session = liveModel.connect()`: It establishes a persistent WebSocket connection to the Gemini model to start a new session. This `LiveSession` object allows for real-time, low-latency streaming of input and output.

3. `session?.send(text)`: The function then sends the initial text prompt to the Gemini Live model through the established session.
4. `session?.startAudioConversation()`: This is a key step that begins the audio part of the conversation. Following connection establishment, the core audio interaction is initiated by calling it. This command signals to the model that the client is ready to begin streaming microphone data.
5. `_liveSessionState.value = LiveSessionState.Running()`: This updates the live session's state to "Running" as soon as the session successfully starts. This value is exposed as a `StateFlow` (backed by `MutableStateFlow`) so that the UI layer can observe it and react in real time.
6. `_liveSessionState.value = LiveSessionState.Error(message = e.localizedMessage)`: Handles errors by updating the `_liveSessionState` to "Error" state and including the relevant message.

Note that, the `@RequiresPermission(Manifest.permission.RECORD_AUDIO)` annotation at the top ensures this function only runs when the **RECORD_AUDIO** permission is granted - necessary for bidirectional audio streaming.

Lifecycle Management: Toggling Session Start/Stop

You learned how to start a session, but you also need to know how to stop the session. The session should be explicitly closed when the microphone is deactivated or when the user navigates away from the screen. Even when you start a new session, the right approach is to stop any ongoing session before starting a new one.

Stopping a session is simple. You can do that by adding this function to `LiveModelManager`:

```
fun stopSession() {
     session?.apply {
       stopAudioConversation()
       stopReceiving()
     }
     _liveSessionState.value = LiveSessionState.Ready()
}
```

The function simply checks if there's a session initiated. If so, it stops the audio conversation with the model.

The `stopReceiving()` method ensures the live session object will no longer receive data from the server, effectively closing the socket connection.

Then it resets the UI state executing `_liveSessionState.value = LiveSessionState.Ready()`. This indicates to the user that there's no ongoing session and you're ready to start a new one.

This will be handy when you tap on the **Mic** button:

1. In the `Ready` state, the app will start a new session.
2. Once the session is `Running`, it will stop the session if you tap the same button.

As `MainViewModel` is responsible for handling user interactions, update the `askAbout()` function in **MainViewModel.kt** to implement the session start/stop toggle:

```
@RequiresPermission(Manifest.permission.RECORD_AUDIO)
fun askAbout(catBreed: String) {
  when (val state = liveSessionState.value) {
    is LiveSessionState.Ready -> {
      liveModelManager.startSessionFromText(catBreed)
    }

    is LiveSessionState.Running -> {
      liveModelManager.stopSession()
    }

    else -> {
      Log.d(TAG, "Live session state: $state")
    }
  }
}
```

The above function starts or stops a live audio session to ask Gemini Live about a specific cat breed. The `catBreed` parameter is the name of the cat breed to ask about. This is used as the initial prompt when starting a new session.

Then it checks the current state of the `liveSessionState` and performs the following:

- If the state is `Ready`, it starts a new live session.
- If the state is `Running`, it stops the currently active session.
- In other states (like `Error` or `Loading`), it logs the current state.

Ready to try this out? **Build and run** the app now. Navigate to the detail screen by selecting a cat breed and then tap the Mic button at the bottom.

You'll see the state change to `RUNNING`, and Gemini Live will start talking about the cat breed you selected!

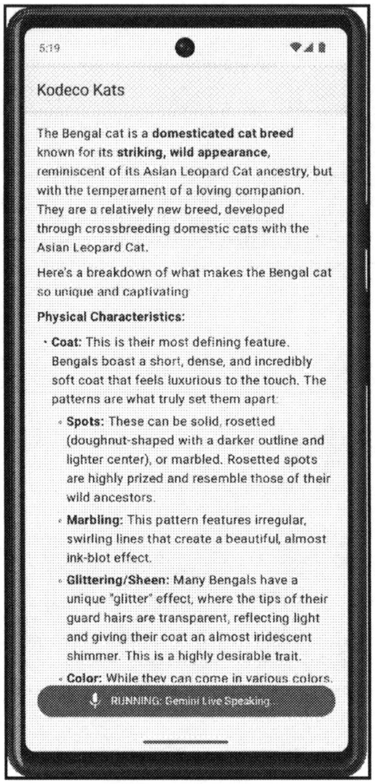

Model Running

Function Calling: Making Gemini Your App's Agent

Now you know how to turn your app into a voice assistant using the Gemini Live API. The next big step is **Function Calling** - the superpower that lets the model actually interact with the logic and functionality of an Android app. It's what makes the voice assistant an **agent** for your app.

Function calling allows the model to determine when an action is necessary to satisfy a user request. The implementation follows a standard multi-step process.

Step 1: Define the App Function and its Declaration

First, you need the actual function in your app that you want the model to be able to call. In the sample app, you may want the user to ask for pictures of a specific cat breed - which means opening a Google Image search.

Here's the function in **LiveModelManager.kt** that handles launching the Google search Intent:

```
fun showPicture(catBreed: String) {
  coroutineScope.launch(Dispatchers.Default) {
    val query = Uri.encode("$catBreed cat pictures")
    val url = "https://www.google.com/search?q=$query&tbm=isch"

    val intent = Intent(Intent.ACTION_VIEW)
    intent.data = Uri.parse(url)
    intent.addFlags(Intent.FLAG_ACTIVITY_NEW_TASK)

    try {
      context.startActivity(intent)
    } catch (e: Exception) {
      Log.e(TAG, "Error opening Google Images", e)
    }
  }
}
```

Next, create a **FunctionDeclaration** to describe this function to the Gemini model. This is like creating an API reference for the model. It needs a name, a plain-English description (which is crucial for the model to understand when to use it), and the required parameters.

```
// The FunctionDeclaration for the model
val showPictureFunctionDeclaration = FunctionDeclaration(
    name = "showPicture",
    description = "Function to show picture of cat breed",
    parameters = mapOf(
        "catBreed" to Schema.string(
            description = "A short string describing the cat
breed to show picture"
        )
    )
)
```

Step 2: Pass the Tool to the LiveModel

The Gemini model needs to know what tools (functions) it has available before the conversation even starts. You need to package the **FunctionDeclaration** into a **Tool** object and pass it to the `liveModel` initialization.

So, define `functionHandlerTool` below `showPictureFunctionDeclaration` as follows:

```
// Packaging the declaration into a Tool
val functionHandlerTool =
Tool.functionDeclarations(listOf(showPictureFunctionDeclaration)
)
```

Then update the `liveModel` initialization inside the `initializeGeminiLive()` function:

```
// Initializing the LiveGenerativeModel
liveModel = Firebase.ai(backend = googleAI()).liveModel(
    modelName = "gemini-live-2.5-flash-preview",
    generationConfig = liveGenerationConfig,
    tools = listOf(functionHandlerTool), // Passing the tool
here!
)
```

Now, the model knows that if a user asks something like "Can you show me pictures of a Siamese cat?", it has a tool named `showPicture` that can handle that request.

Step 3: Implement the Handler Function

When the user says something that triggers the function, the model sends a `FunctionCallPart` to the app. You need a special function — a **handler**, to intercept this call, execute the app logic, and send the result back to the model.

Below `functionHandlerTool`, add the following function to act as the handler:

```
fun functionCallHandler(functionCall: FunctionCallPart):
FunctionResponsePart {
  return when (functionCall.name) {
    "showPicture" -> {
      val catBreed =
functionCall.args["catBreed"]!!.jsonPrimitive.content
      showPicture(catBreed = catBreed)
      val response = JsonObject(
        mapOf(
          "success" to JsonPrimitive(true),
          "message" to JsonPrimitive("Showing pictures of
$catBreed")
        )
      )
      FunctionResponsePart(functionCall.name, response)
    }

    else -> {
      val response = JsonObject(
```

```
            mapOf(
                "error" to JsonPrimitive("Unknown function: $
{functionCall.name}")
            )
        )
        FunctionResponsePart(functionCall.name, response)
      }
    }
}
```

The crucial parts here are:

1. **Extracting the parameters:** The model sends the arguments it inferred from the user's speech (`catBreed` in this case).
2. **Calling the local app logic:** This is where `showPicture()` is executed, opening the browser.
3. **Sending the response back:** You return a `FunctionResponsePart` confirming the execution. The model uses this confirmation to finalize its spoken reply (e.g., "I'm on it! Showing pictures of Siamese cats now").

Step 4: Start the Conversation with a Function Handler

Finally, when you start or continue the live session, pass the handler function to the `startAudioConversation()` call. This tells the `LiveSession` which function to invoke when the model decides to use a tool.

Update the `startSessionFromText()` function as follows:

```
// Start the conversation
session = liveModel.connect()
session?.send(text)
session?.startAudioConversation(::functionCallHandler) // Pass
the handler here!
```

That's it! **Build and run** the app, and ask about your selected cat breed using Gemini Live.

You'll be amazed to hear Gemini ending the description with follow-up questions like: "... Do you want me to show pictures of Siamese cats?"

Reply something affirmative, like "Yes, I'm curious!"

Then the magic happens – Gemini Live handles all the real-time voice streaming and the function calling handshake, resulting in the image search opening right on your phone.

Function Calling

Conclusion

To wrap this up, what you've done with the Gemini Live API and Function Calling isn't just an evolutionary step; it's a massive leap forward in how we build mobile AI experiences.

The chapter started with the core idea: using the Gemini Live API to achieve low-latency, real-time voice streaming without needing a complex backend. That alone moves us beyond the clunky "wait-and-reply" experience of older chatbots. Setting up a "live" model that can manage a live session made real-time interaction possible!

But **Function Calling** is where the true power of an interactive app shines. You've turned Gemini into a genuine agent for your app just by taking just a few steps:

- Defining a function.
- Packaging it as a tool to be used with a model.
- Implementing a simple function call handler.

Now, when a user says, "Show me pictures of a Maine Coon cat," it's not just a conversation; it's a seamless command that triggers native Android code, opening the search Intent directly!

This ability to blend natural language with your app's specific logic is what truly unlocks next-generation, hands-free voice control on Android. It's time to stop building apps that just *talk* - and start building apps that *act*.

Chapter 9: Best Practices, Ethics, and the Future of Android AI

Zahidur Rahman Faisal

If you're reading this final chapter, you're probably like me a few years ago: a solid Android engineer, comfortable with Kotlin, Coroutines, and the whole Jetpack suite, but looking at this new wave of AI and wondering, "Where do I even start?" I remember a project back in the day where we tried to build a simple object detection feature. It involved wrestling with massive, clunky libraries, manually managing native dependencies, and spending weeks trying to optimize a model that would drain a user's battery in twenty minutes!

Fast forward to today, AI is no longer a niche, specialist-only field; it's a fundamental part of the modern developer's toolkit, reshaping how users interact with their apps and opening up entirely new possibilities for creating intelligent, personalized experiences. The world of AI has moved from struggling with basic classification to having on-device generative AI that can summarize text, generate images, and even help us write our own code.

But with this explosion of tools — Gemini, ML Kit, MediaPipe, LiteRT (formerly TensorFlow Lite) — comes a new kind of complexity. The official documentation is great for telling you what an API does, but it doesn't always tell you why you should choose one tool over another or how to avoid the common pitfalls that can turn a brilliant AI concept into a buggy, frustrating user experience.

That's the goal of this book — this isn't just a rehash of the docs. These are the lessons I wish I'd had when I was starting out. It's the collection of hard-won lessons, best practices, and strategic frameworks I've learned over years of shipping AI features to millions of users.

This chapter covers the three crucial stages of building with AI on Android:

1. **The Big Decision:** Start with the single most important architectural question you'll face: Should your AI run on the user's device or in the cloud? This choice impacts everything that follows.
2. **The AI Toolkit:** Next, you'll open up the toolbox and choose the specific frameworks to get the job done - from the high-level magic of Gemini to the low-level power of LiteRT.
3. **Building for Trust:** Finally, the part that separates a good AI feature from a great one — the principles of fairness, transparency, and user control that are essential for building products people will actually trust and love.

The Big Decision: Where Does the "Thinking" Happen?

Before you write a single line of AI-specific code, before you even think about which model to use, you have to answer one fundamental architectural question:

"Where will the AI model perform its inference?"

Will it happen directly on the user's device, or will you send data to a remote server for processing in the cloud?

This isn't a minor implementation detail. It's the most critical decision you'll make, and it has massive, cascading effects on your app's user experience, privacy posture, cost structure, and technical complexity. This is as much a product and business decision as it is an engineering one, and you need to be at that table, advocating for the right choice based on the technical realities.

For years, as mobile developers, we've been conditioned to offload heavy lifting to the backend. Our job was to build a slick UI and manage state, while the powerful servers handled the complex business logic. The rise of powerful on-device AI turns that model on its head. It represents a genuine paradigm shift for us. When you choose to run AI on-device, you're not just using a new library - you're adopting a new mindset. Suddenly, you have to think like an embedded-systems engineer again.

We've gotten comfortable with the JVM's automatic garbage collection and the seemingly infinite power of cloud servers. On-device AI forces us back to first principles. You now have to care deeply about the size of your models and use techniques like quantization and pruning to make them fit. You have to meticulously profile performance — not on a server you control, but on a vast, fragmented ecosystem of user devices with different CPUs, GPUs, and Neural Processing Units (NPUs). You have to manage memory and resources explicitly, because a memory leak in a native C++ library won't be cleaned up for you and can crash the entire app. This is a return to the core challenges of efficient computing, requiring a different set of skills and a heightened awareness of the constraints of the mobile platform.

Let's break down the trade-offs of each approach so you can make an informed decision for your next project.

On-Device AI: The Pros and Cons of Local Intelligence

Running AI models directly on the user's phone is the direction the industry is heading for a wide range of use cases — and for good reason. ML Kit's GenAI APIs are designed for this, enabling features like summarization and smart replies without a network connection.

The Wins

- **Privacy and Security:** When the model runs locally, the user's data — whether photos, messages, or voice recordings — never leaves their device. This is a game-changer for apps in sensitive domains like healthcare, finance, or those aimed at children. Even for general-purpose apps, it's a massive trust-builder. You're not just telling users you respect their privacy; you're proving it through your architecture.

- **Latency & Responsiveness:** It's instantaneous. There is no network round-trip. No waiting for a request to travel to a server, get processed, and come back. This ultra-low latency is what makes features like real-time AR filters, live translation, or motion detection feel snappy and magical. The AI becomes a seamless part of the user experience, not a feature you have to wait for.

- **Offline Functionality:** The user could be on a plane, in a subway tunnel, or hiking in a remote area with no signal. Your AI feature will still function perfectly. This reliability is a huge UX win and can be a key differentiator for apps like travel tools or offline translators.
- **Cost:** No scaling cloud bills. Every API call to a cloud-based AI service costs money. With on-device inference, there are no per-use charges. This can be a massive financial advantage, especially for free apps with large user bases or startups with thin margins. You can scale to millions of users without a proportional increase in your cloud expenses.

The Trade-offs You Accept

- **Hardware Constraints & Model Size:** A smartphone, no matter how powerful, is not a cloud data center. You're fundamentally limited by the device's processor (CPU, GPU, NPU) and available RAM. You simply cannot run the massive, state-of-the-art models available in the cloud. This means you must invest significant effort in model optimization, using techniques like quantization to reduce precision and pruning to remove unnecessary parameters before deployment.
- **Battery Drain:** Complex computations consume power. An inefficiently implemented on-device model can be a major battery hog — a cardinal sin in mobile development. You need to carefully profile your model's energy consumption across a range of devices and optimize accordingly.
- **Update Complexity:** When your model is bundled inside your app, updating it usually means shipping a full app update through the Play Store. This is a much slower iteration cycle than simply deploying a new endpoint to your server. While new services like Play for On-device AI aim to solve this by allowing dynamic delivery of models, it's still a more complex workflow to manage than a typical backend deployment.
- **Development Complexity:** You're now responsible for ensuring your AI feature performs well across the incredibly fragmented Android ecosystem. A model that runs great on a flagship Pixel with a Tensor chip might crawl on a mid-range device with a different chipset. This requires extensive testing and, potentially, creating different versions of your model for different hardware profiles.

Cloud AI: When You Need the Heavy Artillery

Despite the powerful trend toward on-device processing, the cloud still has a critical role to play, especially when you need raw, unadulterated power.

Why You'd Choose It

- **Power & Scalability:** Access to the biggest and best models. When your use case demands deep reasoning, the highest-quality image generation, or the analysis of massive datasets, the cloud is your best option. You get access to advanced models like Gemini Pro or Imagen 4, which are orders of magnitude more powerful than anything that can run on a phone.
- **Ease of Updates:** Iterate at the speed of the web. You can tweak, update, or completely swap out your AI model on the backend at any time, and the changes are live for all users instantly. You don't need to wait for app reviews or user updates. This allows for incredibly rapid experimentation and improvement.
- **Centralized Data Processing & Learning:** If your feature relies on learning from the collective behavior of all your users — think of a recommendation engine like Netflix's or a fraud-detection system — you need a central place to process that data. The cloud is purpose-built for this kind of large-scale aggregation and model training.

The Trade-offs

- **Latency:** The network is your bottleneck. Even on a fast 5G connection, there will always be a noticeable delay as data travels to the server and back. For any feature that requires a real-time feel, this latency can be a deal-breaker.
- **Connectivity Dependence:** No internet, no feature. This is the most obvious drawback. If your user is offline, your AI feature stops working — making it unsuitable for functionality that's core to your app's experience.
- **Privacy Concerns:** You're handling user data. The moment you send data to your server, you become a custodian of that user's information. You must be extremely transparent about this in your privacy policy and implement robust security measures. For many users, this is a significant barrier to trust.
- **Cost:** Pay-as-you-go can get expensive. You're typically charged per API call or per unit of compute time. Manageable at a small scale, yes — but costs can quickly spiral as your user base grows. Monitor this carefully and ensure your business model supports it.

The Pragmatic Engineer's Choice: The Hybrid Approach

After looking at these pros and cons, you might realize that for many sophisticated applications, the answer isn't a strict "either/or." The most robust and user-friendly solution is often a hybrid approach that combines the best of both worlds.

This is the strategy I recommend for most complex projects. You use on-device AI as your "first line of defense." It handles the tasks that need to be fast, private, and always available. This could be things like keyword spotting to activate a feature, pre-processing images to detect faces before uploading, or providing quick, simple text summaries.

Then, when the user needs more power, or the on-device model can't handle the request, your app can gracefully fall back to a more powerful cloud-based model.

This hybrid model gives the user the best possible experience: the speed and privacy of on-device AI for everyday tasks, and the raw power of cloud AI for the heavy-duty features, all managed seamlessly within a single application.

To help you navigate this crucial decision, here's a table that summarizes the key factors. Keep this handy as you plan your next AI feature:

Factor	On-device AI	Cloud AI	When to choose it
Privacy	High (data never leaves the device)	Lower (data sent to servers)	Choose **On-Device** when handling any sensitive user data (health, finance, private messages, or photos).
Offline Use	Fully functional without an internet connection	Requires a stable internet connection	Choose **On-Device** for any core functionality that must remain reliable regardless of connectivity.
Cost Model	No per-inference cost; one-time development cost	Usage-based (per API call or compute time)	Choose **On-Device** if you have a large user base and a business model that can't support scaling API costs.
Model Complexity or Power	Limited by device hardware (smaller models)	Virtually unlimited (access to state-of-the-art models)	Choose **Cloud** for tasks requiring deep reasoning, high-quality generation, or complex analysis.
Update Agility	Slower (requires an app update or model delivery service)	Instant (update the model on the server)	Choose **Cloud** when you need to iterate and improve your model frequently and rapidly.
Device Impact	Higher (consumes battery, storage, and RAM)	Lower (minimal impact on device resources)	Choose **Cloud** if your app is already resource-intensive or targets lower-end devices.

Android AI Toolkit

Alright, you've made the big architectural decision about where the AI will run. Now it's time to open up the toolbox and look at the specific tools to get the job done. The Android AI ecosystem is rich and varied, but it can also be confusing. The key is the "right tool for the job" philosophy. Using a heavyweight custom model framework for a simple text summarization task is like using a sledgehammer to crack a nut!

This table breaks down the main tools in the Android AI stack and what they're best used for. Refer back to this as you're implementing your AI feature:

Tool/API	Primary Use Case	Customization Level	Ease of Use	On-Device/ Cloud
Gemini in Android Studio	AI-powered coding assistant for developers	N/A	Very Easy	N/A
ML Kit GenAI APIs	Adding common on-device generative AI features (e.g., summarization, translation)	Low (Fine-tuned for specific tasks)	Very Easy	On-Device
Firebase AI	Accessing powerful, cloud-based generative AI models	High (Full prompting capabilities)	Easy	Cloud
MediaPipe Solutions	High-performance on-device vision, audio, and text tasks	Medium (Customizable tasks and models)	Medium	On-Device
Custom Models with LiteRT	Running your own custom-trained TensorFlow Lite models	Very High (Full control over the model and runtime)	Hard	On-Device

AI-powered Programming: Gemini in Android Studio

Android Studio is the tool that will help you build everything else. Gemini in Android Studio is your AI-powered pair programmer. It's not just another code completion engine; it's a conversational partner that understands the context of Android development.

Gemini in Android Studio truly shines in its deep integration with the Android ecosystem. This is its superpower compared to a generic tool like ChatGPT. It can help you:

- **Fix Cryptic Gradle Errors:** Just paste that wall of red text from your build output and ask what's wrong. It has been trained on countless Android build issues and can often spot a misplaced comma or version mismatch you've been stuck on for hours.
- **Analyze Crash Reports:** It integrates with Logcat and App Quality Insights. You can literally ask, "Why is my app crashing with this stack trace?" and it will analyze the report and suggest potential causes and fixes.
- **Mock Up and Troubleshoot Compose UIs:** You can describe a UI in plain English, and it will generate the Compose code for you. It's also great for debugging layout issues.

To get the most out of it, you need to learn how to "speak its language." Don't treat it like a search engine.

Mastering Prompts: Getting What You Want Done

Whether you're using Gemini in Android Studio or calling the API from your app, the quality of your output is directly proportional to the quality of your input, or "prompt." Prompt design is a skill, but it's one you can learn.

Here are some practical tips for writing effective prompts:

Be Hyper-Specific with Your Prompts

This is the golden rule. A vague question gets a vague answer. Instead of asking, "How do I use the camera?" ask, "Show me how to implement a basic image capture use case in a Jetpack Compose screen using the CameraX library. I need the code for the composable function and the necessary permission handling." The more context you provide, the better your results will be.

Define the Structure and the Output

Don't just throw a long block of query or prompt at the model; use clear, specific instructions. Add context that the model needs to solve the problem effectively. Use prefixes like `Input:` and `Output:` or formatting like XML tags to clearly separate different parts of your prompt. This helps the model understand the task and the desired format.

Want your code generated in a specific way? Tell it. For example: "Refactor this function to use Kotlin coroutines. Make sure the network call happens on the IO dispatcher and the result is updated on the main thread. Add KDoc comments explaining each parameter." This level of instruction yields far better results than just saying "make this asynchronous".

Break Down Complex Problems

Don't try to solve a complex, multi-step problem in a single prompt. Break the problem down into a sequence of simpler tasks. Make the output of the first prompt the input for the second, and so on.

For example, if you need to build a complex feature, don't ask for the whole thing at once. Ask to create the data model first, then the repository, then the ViewModel, and finally the UI components to display the data.

By breaking the request into a series of simpler, logical steps, you guide the model and get a more comprehensive and accurate result overall.

Building AI That People Actually Trust

Now you know the architecture and the tools, you can build a technically functional AI feature, but the job isn't done. Technical implementation is only half the battle. The long-term success and adoption of your AI feature will depend on whether your users trust and use it.

This section isn't about optional, "nice-to-have" features. This is about our core responsibilities as engineers when we build systems that make decisions, generate content, and interact with users in increasingly human-like ways.

When you read corporate blog posts about "Responsible AI," it's easy to dismiss them as vague, high-level marketing talk. But when you dig into the details, you realize that these principles translate into concrete, actionable engineering tasks. Responsible AI is not an abstract ethical guideline; it is a multi-layered engineering discipline that needs to be designed and implemented with the same rigor as security or performance testing.

This discipline spans the entire stack. It starts with the **data layer**, where you must be conscious of mitigating the biases in the data used to train your models. It extends to the **application layer**, where you are responsible for implementing secure data handling, writing clear privacy policies. It manifests in the **UI/UX layer**, where your job is to make the AI's work visible, explain its decisions in simple terms, and give users meaningful control. And it even connects to the **platform layer**, where you must respect the global settings users have for managing AI features and permissions across their devices.

Designing for Fairness: How to Avoid Building Biased Bots

First, let's define "fairness" in a practical way that we, as engineers, can work with. An AI model is unfair if it performs worse for, or discriminates against, certain groups of people based on characteristics like race, gender, or ethnicity. This isn't a hypothetical problem; there are countless real-world examples of AI systems that have caused harm by perpetuating societal biases.

1. **It Starts with the Data:** The primary source of bias in AI is the data it was trained on. If a model was trained on a dataset where most of the pictures of doctors were men, it might be less likely to correctly identify a woman in a photo as a "doctor". While we as app developers don't always train the models we use (especially when using pre-trained models from external sources), we still have a responsibility to be aware of this potential and test for it in the context of our app.

2. **Proactive Testing and Monitoring:** You cannot assume a model is fair. You must verify it. This means actively testing your AI features with a diverse range of inputs and user groups. Does your image recognition feature work as well for people with darker skin tones? Does your voice transcription handle different accents equally well? You need to build a framework for continuously monitoring the performance of your model across different user segments and develop a plan to address any disparities you find.

Putting Users in Control: The Non-Negotiable Settings

Giving users clear, accessible controls is a fundamental requirement for building an ethical and trustworthy application. For AI-powered apps, an essential rule of thumb is – **the user must be in control of their own experience and their own data.**

These are the essential controls you should build into any app with AI features:

- **A Clear Opt-Out:** In your app's settings, there should be a simple, easy-to-find toggle switch for each major AI feature, as well as optionally a master switch to disable all of them. The user should never feel like they are being forced to use an AI feature they don't want.
- **Granular and Contextual Permissions:** Follow the platform's best practices for permissions. Don't ask for access to the camera, microphone, or location when the user first launches the app. Instead, request each permission contextually, at the moment the feature needs it, and provide a clear explanation of why you need it. For example, when the user taps the voice input button, that's the time to request microphone permission with a dialog that says, "Allow this app to access your microphone to enable voice commands".
- **Data Management and Deletion:** If your AI uses the user's data to provide a personalized experience, you should give them a way to view and manage that data. A "data dashboard" where a user can see what information the app has collected and easily delete their history is a powerful tool for building transparency and control.
- **Respect System-Level Settings:** Remember that users can disable AI features at the operating system level (for example, by turning off Google Assistant or revoking your app's permissions in the system settings). Your app needs to detect these system-level changes and react gracefully, disabling the relevant features without crashing or showing a broken UI.

Conclusion

If you've made it to the end of this chapter, then you already understand something many developers never quite grasp: building AI features on Android isn't just about gluing a model onto an app. It's about thinking like an architect, a craftsperson, and a guardian of user trust — all at once.

By now, you've learned that the very first question you face isn't *"Which model should I use?"* but *"Where should the thinking happen?"* That single decision: on-device, cloud, or a hybrid of both, shapes everything that follows. Pick the right path and the pieces fall into place. Pick the wrong one and you'll be wrestling your own architecture for months.

You also explored the expanding toolbox Google has placed in your hands: **Gemini** in Android Studio, **ML Kit**'s on-device generative APIs, **MediaPipe**'s high-performance pipelines, **Firebase AI** for cloud power, and the raw, low-level control of custom **LiteRT** models. The real skill is knowing when to use what to get things done.

But perhaps the biggest shift is the rise of prompting as a core engineering skill. Whether you're guiding Gemini in Android Studio to churn out clean Compose code or structuring precise few-shot prompts for a cloud model, you're no longer just writing code... you're steering intelligence. And like any craft, the more intentional your inputs, the more reliable the outputs.

Still, none of the technical mastery matters if your users can't trust what you build. The future of Android AI won't be written by the developers who build the flashiest demos — it'll be written by the ones who build responsibly. Responsible AI is no longer optional, it's engineering.

You're stepping into an ecosystem that is evolving at breakneck speed. Models will change. APIs will come and go. But the principles you've learned here — architectural clarity, tool literacy, careful prompting, and ethical responsibility - will stay relevant, no matter how powerful the next models become.

So if there's one message to carry with you from this chapter, it's this:

AI isn't replacing Android developers. It's amplifying the ones who adapt.

Thank you for reading till the very end... I'm wishing you an exciting journey into the world of AI & Android ahead. Your journey as an AI engineer on Android is just beginning, and I, for one, can't wait to see what you build!

10 Conclusion

What a journey it's been! From exploring the AI-powered Android landscape to diving deep into on-device intelligence with ML Kit and custom ML with MediaPipe, to harnessing the power of the cloud with Firebase and Gemini Live, you've gotten a taste of building intelligent Android applications with advanced AI and ML techniques.

This book isn't meant to be the final word on Android AI; it's a solid foundation. The most important thing you can do now is start building. Experiment with these tools. Make mistakes. Profile your performance. And most importantly, listen to your users.

If there are 3 things to take away from this book, they're these:

1. **Start with the "where".** The on-device vs. cloud decision is your foundational choice. Get it right, and everything else becomes easier. Get it wrong, and you'll be fighting against your own architecture for the entire lifecycle of the project.
2. **Choose the right tool for the job.** Don't reach for the most complex tool when a simpler one will do. Start with high-level APIs like ML Kit and MediaPipe. Only move down into the depth and complexity of a custom LiteRT implementation when you truly need that level of control.
3. **Build for trust from day one.** Fairness, transparency, and user control are not afterthoughts. They are core engineering requirements. Integrate them into your design process, user interface, and application logic right from the start.

We hope the concepts you learned in this book inspire you to explore and experiment with different AI and ML practices, and to create smarter, more interactive apps! Your journey as an AI engineer on Android is just beginning, and I, for one, can't wait to see what you build.

If you're hungry for more, we recommend exploring the official Android documentation for AI and ML Kit, and diving deeper into the capabilities of Gemini and other generative AI models.

If you have any questions or comments as you work through this book, please stop by our forums at https://forums.kodeco.com and look for the particular forum category for this book.

Thank you again for purchasing this book. Your continued support is what makes the books, tutorials, videos and other things we do at kodeco possible. We truly appreciate it!

– The Android AI & ML team

Made in the USA
Monee, IL
05 November 2026

67811059R00103